PRACTICES FOR INTERIM STORAGE OF RESEARCH REACTOR SPENT NUCLEAR FUEL

The following States are Members of the International Atomic Energy Agency:

AFGHANISTAN
ALBANIA
ALGERIA
ANGOLA
ANTIGUA AND BARBUDA
ARGENTINA
ARMENIA
AUSTRALIA
AUSTRIA
AZERBAIJAN
BAHAMAS
BAHRAIN
BANGLADESH
BARBADOS
BELARUS
BELGIUM
BELIZE
BENIN
BOLIVIA, PLURINATIONAL STATE OF
BOSNIA AND HERZEGOVINA
BOTSWANA
BRAZIL
BRUNEI DARUSSALAM
BULGARIA
BURKINA FASO
BURUNDI
CAMBODIA
CAMEROON
CANADA
CENTRAL AFRICAN REPUBLIC
CHAD
CHILE
CHINA
COLOMBIA
COMOROS
CONGO
COSTA RICA
CÔTE D'IVOIRE
CROATIA
CUBA
CYPRUS
CZECH REPUBLIC
DEMOCRATIC REPUBLIC OF THE CONGO
DENMARK
DJIBOUTI
DOMINICA
DOMINICAN REPUBLIC
ECUADOR
EGYPT
EL SALVADOR
ERITREA
ESTONIA
ESWATINI
ETHIOPIA
FIJI
FINLAND
FRANCE
GABON
GEORGIA
GERMANY
GHANA
GREECE
GRENADA
GUATEMALA
GUYANA
HAITI
HOLY SEE
HONDURAS
HUNGARY
ICELAND
INDIA
INDONESIA
IRAN, ISLAMIC REPUBLIC OF
IRAQ
IRELAND
ISRAEL
ITALY
JAMAICA
JAPAN
JORDAN
KAZAKHSTAN
KENYA
KOREA, REPUBLIC OF
KUWAIT
KYRGYZSTAN
LAO PEOPLE'S DEMOCRATIC REPUBLIC
LATVIA
LEBANON
LESOTHO
LIBERIA
LIBYA
LIECHTENSTEIN
LITHUANIA
LUXEMBOURG
MADAGASCAR
MALAWI
MALAYSIA
MALI
MALTA
MARSHALL ISLANDS
MAURITANIA
MAURITIUS
MEXICO
MONACO
MONGOLIA
MONTENEGRO
MOROCCO
MOZAMBIQUE
MYANMAR
NAMIBIA
NEPAL
NETHERLANDS
NEW ZEALAND
NICARAGUA
NIGER
NIGERIA
NORTH MACEDONIA
NORWAY
OMAN
PAKISTAN
PALAU
PANAMA
PAPUA NEW GUINEA
PARAGUAY
PERU
PHILIPPINES
POLAND
PORTUGAL
QATAR
REPUBLIC OF MOLDOVA
ROMANIA
RUSSIAN FEDERATION
RWANDA
SAINT KITTS AND NEVIS
SAINT LUCIA
SAINT VINCENT AND THE GRENADINES
SAMOA
SAN MARINO
SAUDI ARABIA
SENEGAL
SERBIA
SEYCHELLES
SIERRA LEONE
SINGAPORE
SLOVAKIA
SLOVENIA
SOUTH AFRICA
SPAIN
SRI LANKA
SUDAN
SWEDEN
SWITZERLAND
SYRIAN ARAB REPUBLIC
TAJIKISTAN
THAILAND
TOGO
TONGA
TRINIDAD AND TOBAGO
TUNISIA
TÜRKİYE
TURKMENISTAN
UGANDA
UKRAINE
UNITED ARAB EMIRATES
UNITED KINGDOM OF GREAT BRITAIN AND NORTHERN IRELAND
UNITED REPUBLIC OF TANZANIA
UNITED STATES OF AMERICA
URUGUAY
UZBEKISTAN
VANUATU
VENEZUELA, BOLIVARIAN REPUBLIC OF
VIET NAM
YEMEN
ZAMBIA
ZIMBABWE

The Agency's Statute was approved on 23 October 1956 by the Conference on the Statute of the IAEA held at United Nations Headquarters, New York; it entered into force on 29 July 1957. The Headquarters of the Agency are situated in Vienna. Its principal objective is "to accelerate and enlarge the contribution of atomic energy to peace, health and prosperity throughout the world".

IAEA NUCLEAR ENERGY SERIES No. NF-T-3.10

PRACTICES FOR INTERIM STORAGE OF RESEARCH REACTOR SPENT NUCLEAR FUEL

INTERNATIONAL ATOMIC ENERGY AGENCY
VIENNA, 2022

Marketing and Sales Unit, Publishing Section
International Atomic Energy Agency
Vienna International Centre
PO Box 100
1400 Vienna, Austria
fax: +43 1 26007 22529
tel.: +43 1 2600 22417
email: sales.publications@iaea.org
www.iaea.org/publications

Printed by the IAEA in Austria
September 2022
STI/PUB/2007

IAEA Library Cataloguing in Publication Data

Names: International Atomic Energy Agency.
Title: Practices for interim storage of research reactor spent nuclear fuel / International Atomic Energy Agency.
Description: Vienna : International Atomic Energy Agency, 2022. | Series: IAEA nuclear energy series, ISSN 1995–7807 ; no. NF-T-3.10 | Includes bibliographical references.
Identifiers: IAEAL 22-01508 | ISBN 978–92–0–123122–2 (paperback : alk. paper) | ISBN 978–92–0–123222–9 (pdf) | ISBN 978–92–0–123322–6 (epub)
Subjects: LCSH: Spent reactor fuels — Storage. | Radioactive wastes — Management. | Reactor fuel reprocessing. | Nuclear reactors — Safety measures.
Classification: UDC 621.039.74 | STI/PUB/2007

FOREWORD

The IAEA's statutory role is to "seek to accelerate and enlarge the contribution of atomic energy to peace, health and prosperity throughout the world". Among other functions, the IAEA is authorized to "foster the exchange of scientific and technical information on peaceful uses of atomic energy". One way this is achieved is through a range of technical publications including the IAEA Nuclear Energy Series.

The IAEA Nuclear Energy Series comprises publications designed to further the use of nuclear technologies in support of sustainable development, to advance nuclear science and technology, catalyse innovation and build capacity to support the existing and expanded use of nuclear power and nuclear science applications. The publications include information covering all policy, technological and management aspects of the definition and implementation of activities involving the peaceful use of nuclear technology.

The IAEA safety standards establish fundamental principles, requirements and recommendations to ensure nuclear safety and serve as a global reference for protecting people and the environment from harmful effects of ionizing radiation.

When IAEA Nuclear Energy Series publications address safety, it is ensured that the IAEA safety standards are referred to as the current boundary conditions for the application of nuclear technology.

In recent years, international activities related to the back end of the nuclear fuel cycle have been focused on spent fuel take-back programmes, namely the Foreign Research Reactor Spent Nuclear Fuel Acceptance Program of the United States of America (USA) and the Russian Research Reactor Fuel Return programme. The objective of these respective take-back programmes for fuels originating in the USA and the Russian Federation is to eliminate inventories of high enriched uranium by returning research reactor spent nuclear fuel (RRSNF) to the country where the fuel was originally enriched. Eventually when these programmes have achieved their objective and research reactors no longer store high enriched uranium and the demand for high enriched uranium for research reactors has decreased, it is almost certain that the take-back programmes will cease operation. Countries with one or more research reactors but no nuclear power programme will either need to manage the relatively small amounts of spent fuel generated in their research reactors or permanently shut down their research reactors before the take-back programmes come to an end.

It is probable that hundreds of research reactors worldwide, both operational and shutdown but not yet decommissioned, will continue storing RRSNF for a long time. As a consequence, safe, secure, reliable and cost effective handling and interim storage of RRSNF will remain crucial issues for Member States with research reactors.

Recognizing these needs, this publication captures valuable experience using wet and dry methods for interim storage of RRSNF that have been used for many decades with excellent results at different sites. Aluminium clad RRSNF has been kept intact in wet storage with properly maintained water quality for more than 60 years. Dry storage of RRSNF is also a reliable technology that, when properly implemented, ensures the long term integrity of RRSNF.

This publication is a collection of Member States' experiences and case studies to help the research reactor community manage their issues related to the extended storage of RRSNF. The publication is also intended to assist decision makers in Member States with operating research reactors to identify solutions for RRSNF storage under the economic and technological realities of their countries and with due consideration of the safety and security concerns usually associated with RRSNF.

The IAEA wishes to thank all the contributors to this publication, in particular J.W. Lian (Canada), M. Verberg (Netherlands), N.C. Iyer and D.W. Vinson (USA). The IAEA officers responsible for this publication were P. Adelfang, S. Tozser, F. Marshall and S. Geupel of the Division of Nuclear Fuel Cycle and Waste Technology.

EDITORIAL NOTE

CONTENTS

1. INTRODUCTION

1.1. BACKGROUND

For over 60 years, research and test reactors have made valuable contributions to the development of nuclear power, basic science, education, training, materials development, and radioisotope production for medicine and industry.

During operation of these research reactors, thousands of fuel elements were, and continue to be used, composed of different designs, types, shapes, material composition and enrichment [1]. With very few exceptions (such as when there is no provision for fuel replacement), when the spent fuel reaches its burnup limit it is removed from the reactor core and replaced with a new fuel element, allowing the reactor to operate for many years.

Typically, after it has been discharged from the reactor core, the research reactor spent nuclear fuel (RRSNF) is placed into wet storage, either in the reactor pool, or in a pool away from the reactor. The fuel is maintained there until the decay heat and radiation levels are low enough to allow the movement of the material to the next step in the management of the fuel. Depending on the fuel consumption and conditions of the facility, the wet storage can be extended for long periods of time, in some cases more than 50 years, or the RRSNF may be transferred to dry storage sites, which may be suitable for longer periods.

Neither wet nor dry storage is intended to be the end point of the research reactor fuel cycle; the term 'storage' implies a temporary situation and that the fuel will be retrieved and moved again. Continued generation and storage of spent fuel without full commitment to a clearly defined end point is not a sustainable policy. The end point of the research reactor fuel cycle is expected to be associated with a geological repository for the disposal of the spent fuel assemblies after their conditioning and/or for the disposal of the vitrified radioactive waste arising from spent fuel reprocessing. In the context of this publication, the term 'interim' is used to emphasize that storage is not the final step in the management of spent fuel from research reactors. For the purpose of this publication, interim storage has been defined as any storage period up to 100 years.

Developing a geological repository for spent nuclear fuel (SNF) and high level waste (HLW) is not a simple undertaking; at present, there is no spent fuel repository in operation in the world. The technology and costs involved for the development and maintenance of a geological repository for SNF make it prohibitive for most States, especially for those with only one or two research reactors and no nuclear power programme. It is likely that, for the foreseeable future, wet and dry storage will be options heavily used for the management of spent fuel from research reactors in most States.

Regardless of how long the storage period is, public concerns about the RRSNF remain. In addition to the proliferation, safety and physical security concerns, the organization operating a research reactor or the government has the responsibility to ensure the safe, secure and economic management of its RRSNF [2].

In 1993, the IAEA organized an Advisory Group Meeting on Storage Experience with Spent Fuel from Research Reactors to identify, discuss and plan activities related to RRSNF storage. For many research reactors, the capacity for spent fuel storage had reached or was close to reaching the design storage limits, raising concerns about the feasibility of expanding storage capacity and, from a materials science point of view, about ageing materials in ageing storage facilities, with the related consequences for the integrity of the fuel elements.

Following the recommendations of the experts, the IAEA initiated a series of programmatic activities to assist organizations operating research reactors in dealing with spent fuel management issues. These activities included workshops, technical meetings and coordinated research projects (CRPs), which resulted in several IAEA publications made available to the research reactor community. Examples of such activities are:

— Exchange of experience in options, procedures and practices for RRSNF storage [1, 3];
— Ageing management of materials in RRSNF storage facilities [4, 5];
— Study of corrosion and other forms of material ageing leading to the degradation of mechanical and physical properties of RRSNF [6, 7];
— Study of regional solutions for research reactors in Latin America [8];
— Exchange of experience in the two international RRSNF take-back programmes: the USA Foreign Research Reactor Spent Nuclear Fuel (FRRSNF) acceptance programme and the Russian Research Reactor Fuel Return (RRRFR) programme [9].

1.1.1. Options for the management of spent nuclear fuel from research reactors

After having been discharged from the reactor core, the RRSNF is usually stored under water for cooling. This typically occurs in at-reactor (AR) facilities for three to five years, in order to allow for radioactive decay of spent fuel fission products and to remove the residual decay heat. Following this, the fuel may be moved to another storage location or moved to a disposal facility. In general, organizations operating a research reactor have the following options for the next step in their spent fuel management programme:

— Continue to store the fuel in wet storage;
— Transfer the fuel to a dry storage facility;
— Send the fuel for reprocessing with a domestic or international commercial service provider;
— Return the fuel to the country where it was originally enriched (in this way transferring the responsibility for disposal to others);
— Dispose of the fuel directly into a national or regional geological repository, with or without conditioning.

Additional information regarding options for the full RRSNF management programme can be found in IAEA Nuclear Energy Series No. NF-T-3.9, Research Reactor Spent Fuel Management: Options and Support to Decision Making [10]. This publication also presents a methodology that can be used by a State's government or organizations operating a research reactor to evaluate the multiple options for selection of a single strategy for spent fuel management. Supporting information about the individual options presented above can be found in the following IAEA publications: IAEA-TECDOC-1593, Return of Research Reactor Spent Fuel to the Country of Origin: Requirements for Technical and Administrative Preparations and National Experiences [9], IAEA-TECDOC-1632, Experience of Shipping Russian-origin Research Reactor Spent Fuel to the Russian Federation [11], IAEA Nuclear Energy Series No. NW-T-1.11, Available Reprocessing and Recycling Services for Research Reactor Spent Nuclear Fuel [12], IAEA-TECDOC-1587, Spent Fuel Reprocessing Options [13], and also in Ref. [14].

1.1.2. Current situation of research reactors and spent fuel storage

According to the IAEA Research Reactor Database, some 818 research reactors[1] have been built in 70 countries since the Chicago graphite pile CP-1 went critical in December 1942. In addition to the 819 reactors (including CP-1) that were constructed, 16 were cancelled, 11 are under construction, and 16 are planned as of March 2021 [15].

The breakdown of the 819 reactors by their operational status reveals that 223 are still operational in 53 Member States, 14 have been temporarily shut down, 71 have been in extended or permanent

[1] Research reactors here include also zero power facilities, subcritical assemblies as well as some prototype reactors.

TABLE 1. OPERATIONAL STATUS OF RESEARCH REACTORS BY REGION (MARCH 2021)

Region	Operational	Extended or permanent shutdown	Temporary shutdown	Decommissioned
Africa	8	3	1	1
Middle East and South Asia	19	3	0	6
South East Asia and Pacific	6	1	0	2
Latin America	16	3	1	3
North America	55	19	0	237
Western Europe	23	17	3	123
Eastern Europe	72	19	3	58
Far East	24	6	6	16
Total	223	71	14	446

shutdown, 65 are being decommissioned and 446 have already been decommissioned. Table 1shows the regional distribution and status of the 841 research reactors listed in the database (excluding those that were cancelled, under construction or planned, or being decommissioned) as of March 2021.

A survey performed by the IAEA in 2006 indicated that most spent fuel was stored in light water filled pools or basins [16]. The result of the survey, illustrated in Fig. 1 on the basis of the data presented in [16], showed that many facilities had an auxiliary away-from-reactor pool or drywell.

IAEA Nuclear Energy Series No. NW-T-1.14 (Rev. 1), Status and Trends in Spent Fuel and Radioactive Waste Management [17], provides global inventory data for spent fuel originating from research reactors and other kinds of non-power reactors (e.g. experimental, prototype or naval propulsion reactors). The regional distribution of the spent fuel inventories in wet and dry storage, respectively, from

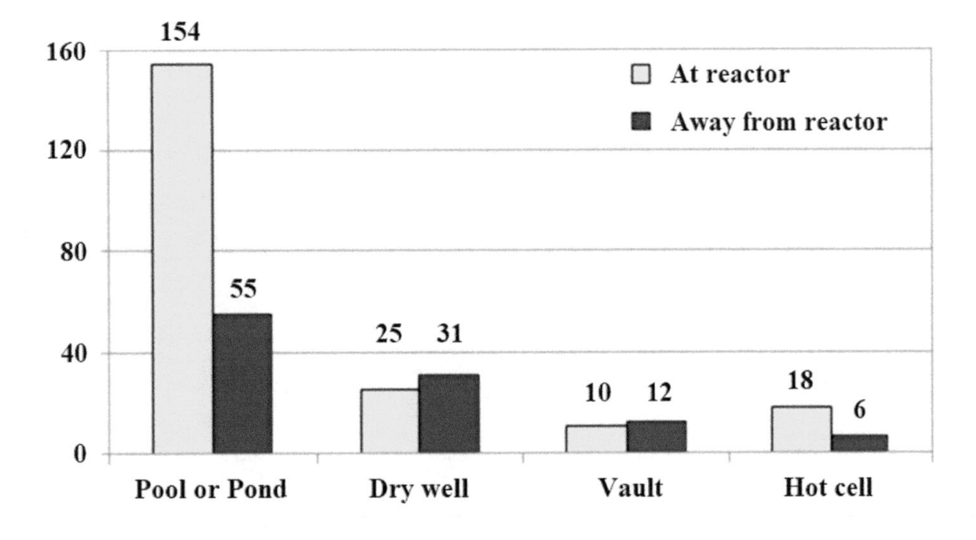

FIG. 1. Number of facilities used for interim storage of RRSNF [16].

these reactors (as of December 2016) is summarized in Table 2. The amounts of spent fuel in Table 2 are given in metric tonnes of heavy metal (t HM) and describe the mass of heavy metals (e.g. uranium, plutonium, thorium and minor actinides) contained in the spent fuel. All countries from which information is available are included in the regional distribution and the global inventory.

It is noteworthy that the global inventory of 4133 t HM arising from research reactors and other kinds of non-power reactors is only 1% of the global inventory of spent fuel which originates from commercial nuclear power plants (NPPs). A typical research reactor has a core capacity in the order of a few kilograms of uranium fuel, compared to 100 tonnes or more for a commercial NPP. The spent fuel quantities from research reactors are generally not publicly reported to the same level of detail as for NPPs, in particular when the reactors are used primarily for military or defence purposes.

From Table 2, it can be gathered that the majority of the spent fuel in storage is in North America. The reason is that spent fuel from prototype power reactors is considered under the research reactor category in Canada and the USA. In many countries, most or all spent fuel from research reactors has been returned to the country of origin (usually the USA or the Russian Federation) for reprocessing or disposal; in these cases, the returned spent fuel will become part of the inventory of the receiving country.

Because most organizations operating a research reactor have few options for the management of spent fuel, storage remains the preferred option while decisions about, and preparations for, geological disposal are undertaken (in the case that options for return of the fuel to the country of origin or shipment of the fuel to a reprocessing facility do not exist). This is why it is expected that many organizations operating a research reactor will extend the operating lifetime of their storage facility, wet or dry, AR or away from the reactor. Therefore, it is important that operating organizations give special consideration to the conditions of the interim storage environment, in order to maintain the integrity of the RRSNF in storage for as long as possible in a safe, secure, reliable and economic condition.

TABLE 2. REPORTED SPENT FUEL INVENTORIES STORED FROM RESEARCH REACTORS AND OTHER NON-POWER REACTORS (AS OF 31 DECEMBER 2016) [17]

Region	Wet storage [t HM]	Dry storage [t HM]	Total [t HM]
Africa	1	1	2
Middle East and South Asia	n.a.	n.a.	n.a.
South East Asia and Pacific	1	n.a.	1
Latin America	1	n.a.	1
North America	39	2920	2959
Western Europe	964	36	1000
Eastern Europe	50	11	61
Far East	109	n.a.	109
Global inventory	1165	2968	4133

Note: n.a.: not applicable (or none reported).

1.2. OBJECTIVES

This publication is intended to help organizations operating research reactors and RRSNF storage facilities identify the most suitable approach for interim storage of their spent fuel. This publication provides:

— Good practices relevant to long term wet and dry storage of RRSNF;
— Experience on lifetime management and lifetime extension issues of the entire storage system and infrastructure relative to RRSNF performance during storage;
— Suggestions for further optimization of effective and safe storage of RRSNF through application of new approaches, such as refinements to the wet storage chemistry envelope.

Much of the information presented in this publication was collected in a meeting of experts, in which they described good practices for RRSNF management [14]. During this meeting, the experts agreed that, while several detailed reports have been issued on the technical basis and lessons learned for wet storage, there is very limited information and guidance provided on dry storage. Therefore, they recommended that this publication describe the main aspects of both options, with an emphasis on dry storage technologies.

1.3. SCOPE

The scope of the publication entails five key elements:

(1) RRSNF fuel types considered for wet storage or dry storage;
(2) Characterization data needs for interim storage: addresses relevant characterization techniques to support prudent RRSNF storage practices;
(3) Wet storage considerations: key aspects of extended wet storage management, including basis and strategies for wet storage of failed RRSNF;
(4) Dry storage considerations;
(5) Lessons learned and current practices for RRSNF wet and dry storage: in terms of the entire storage cycle or discrete portions of the storage life cycle, such as drying treatment, surveillance programme and characterization.

1.4. STRUCTURE

The present publication includes eight sections:

Section 1 provides background information on the RRSNF management challenges and outlines the objective and the scope of this publication. Section 2 presents current wet storage options and considerations. Section 3 presents the challenges and practices for transitioning the RRSNF from wet storage to dry storage. Section 4 presents the current RRSNF dry storage technology. Section 5 presents the considerations for characterizing RRSNF prior to placing it into storage. Section 6 discusses safety issues of RRSNF interim storage. Section 7 presents some examples of wet and dry storage experiences from Member States. Section 8 presents conclusions, drawn during the development of this publication.

2. WET STORAGE OF RESEARCH REACTOR SPENT FUEL

Except for critical assemblies, in which fuel burnup is very low and allows handling of the fuel element in air sometime after the reactor shutdown, most RRSNF in research and test reactors is stored under water for decay and cooling, typically in the reactor pool or in a service pool close to the reactor pool. The water in the selected spent fuel storage pool provides: (a) the necessary shielding for the radiation generated by fuel decay; and (b) the conductive and convective transfer of the heat generated by decay. Typically, the spent fuel is kept in this environment for a period of three to five years. During this period, research reactor operating organizations need to plan for the next step in the RRSNF management strategy.

As discussed in Section 1.1, a commonly used option is to keep the spent fuel in the same pool used for decay and residual heat removal, or to transfer it to a pool especially designed for interim storage. A survey by the IAEA shows that, during the last decades, this has been the preferred option of research reactor operating organizations [16].

Storage of spent fuel in spent fuel pools is a relatively mature technology, and this section provides a review of interim wet storage technology and practices that provide for optimum interim spent fuel storage within spent fuel pools. There are more than six decades of experience in the operation and maintenance of spent fuel pools. The information within this section draws on the collective international knowledge and experience to provide practical direction for operating and maintaining spent fuel pools for safe interim wet storage prior to transition to dry storage or disposal.

2.1. WET STORAGE TECHNOLOGY

Wet storage is implemented in facilities in which spent fuel is kept under water following discharge from the reactor in storage pools. The storage pool is a reinforced concrete structure usually built above ground or at ground elevation. Some early pools were open to the atmosphere, but operational experience and the need to control pool water purity recommend that, for the purpose of spent fuel storage, the pool has to be covered. Some national regulators require the reinforced concrete structure of the pool, including the covering building, to be seismically qualified.

Most pools are stainless steel lined, although some are coated with epoxy resin based paint. However, there has been experience with degradation of the latter after several years. While not recommended, another option is for the pool to be unlined and untreated. Additionally, the pool could be lined with stainless steel or epoxy up to the water line.

Usually the pools are filled with deionized water, with or without additives depending on the type of fuel to be stored and the adopted method of water treatment. The water is either a fixed quantity or a once through pool purge. Water activity levels are maintained as low as reasonably achievable (known as ALARA) by either in-pool or external ion exchange systems or by limiting activity release to the bulk pool water.

2.1.1. Corrosion in wet storage

Water is a strong polar solvent; for this reason, metallic materials tend to corrode in water and water based solutions, presenting a challenge for the storage of spent fuel.

Research reactor fuel is composed of many metallic materials. Aluminium and aluminium alloys are used in a large fraction of research reactor fuel alloy and cladding material. Stainless steels and zirconium alloys are also used for fuel cladding material. Because corrosion control limits such as pH, chloride ion and other impurities are more restrictive for aluminium than for stainless steel and zirconium alloys, the consideration of corrosion in spent fuel storage basins is limited to aluminium and aluminium alloys.

Corrosion can be minimized by storing the spent fuel in monitored water of good quality. A robust treatment on the challenges of wet storage can be found in IAEA Nuclear Energy Series No. NP-T-5.2, Good Practices for Water Quality Management in Research Reactors and Spent Fuel Storage Facilities [18].

2.1.2. Performance of spent nuclear fuel and lifetime prediction models

A general prediction of fuel performance can be made for fuel stored in very good quality water that is consistent with the water quality limits discussed in NP-T-5.2 [18]. Research reactor spent fuel cladding under these conditions may experience localized attack of ~0.0254 mm/year ((1 mil/year) general corrosion rates are much lower), and eventually localized attack at the fuel-plate–side-plate crevice could ultimately lead to the side plate dislodging from the fuel and render retrievability difficult.

At the Savannah River Site (SRS) in the USA, aluminium clad SNF has been wet stored in very good water quality for nearly five decades and has shown no incidence of corrosion attack. Based on this experience and on the results of the ongoing SRS corrosion surveillance programme, it was determined that localized attack is possible even in good quality water, but that a period of at least 50 to 100 years may be needed to initiate that localized attack (to concentrate aggressive species in crevice regions of fuel).

2.2. WET STORAGE FACILITY CONSIDERATIONS

In general, the RRSNF wet storage facility consists of a large pool, with good quality water, where the spent fuel is kept in a subcritical configuration. The pool is a structure made of reinforced concrete, specially designed to serve as biological shielding and ensure the integrity of the spent fuel.

The choice of pool components and materials is dependent upon the type of fuel being stored, cost and facilitation for final pool decommissioning. The latter property has come to the fore as experience in both in-pool performance and final pool decommissioning has been gained.

In most cases, the storage pool was designed as an extension of the research reactor pool, intended to have a lifetime similar to the reactor. Many organizations operating a research reactor have established a continuous refurbishment and modernization programme for their installations as part of a lifetime extension programme and to comply with present safety standards. Therefore, the lifetime extension programme has to include the storage pool. Sections 2.2.1 to 2.2.3 present a brief discussion about concrete, materials used to avoid contact between concrete and water, the ageing mechanisms that may cause degradation of pool and basin structures, and components affecting efficacy of the components related to long term storage in a spent fuel storage pool. Design considerations for surveillance, maintenance and repair within the storage pool are briefly discussed in Section 2.2.4.

2.2.1. Storage pool structure

The structural integrity of a system, structure or component of a mechanical system is typically defined according to its leak tightness and structural stability. Concrete is the most important structural material used in a spent fuel storage pool design and construction. Pools are concrete structures with components such as metallic and non-metallic liners, waterproof materials, joint sealants, and protective coatings and membranes to protect concrete from aggressive fluid and other environmental exposure.

The primary safety functions of concrete spent fuel storage pools are to provide a leak barrier and serve as radiation shielding, while maintaining structural integrity against large failure for design basis loads. The structural integrity of the concrete is essential to continued safe storage of spent fuel in pools, and activities to monitor and inspect basins (e.g. core sampling) are recommended to ensure that these functions are maintained throughout the desired service life of the basin.

Concrete structures can last hundreds of years if maintained and protected from mechanisms that degrade concrete. However, due to the various spent fuel pool designs, the absence or failure of an effective liner, and working and environmental conditions, the structural concrete of the fuel pool may experience one or more degradation mechanisms. Degradation mechanisms of concrete structures include chemical corrosion, temperature extremes, leaching of calcium and silicon, carbonation and corrosion of the reinforcement steel bar, irradiation induced loss of compressive strength, and physical wear.

A good discussion of the aspects of concrete, its formulation and material properties, as well as ageing topics, is provided in Technical Reports Series No. 443, Understanding and Managing Ageing of Material in Spent Fuel Storage Facilities [6], and in Ref. [19]. For a discussion of concrete degradation mechanisms, see NP-T-5.2 [18].

2.2.2. Pool liners and coatings

Two methods may be considered for lining the spent fuel storage pools. The first method is to line the pool with welded sheets of stainless steel or aluminium. The second method is to coat the concrete substrate with a continuous and impervious epoxy resin. Most light water reactor AR storage pools are lined with stainless steel, while both stainless steel and epoxy liners are used for Canada deuterium–uranium (known as CANDU) reactors AR storage. Almost all AFR storage installations are stainless steel lined (France, Finland, Japan, the Russian Federation, Slovakia, Sweden), with the exception of the United Kingdom AFR storage pools which use epoxy coated concrete.

Both of these methods can provide favourable long term performance. In concrete pools coated with epoxy, the concrete employed has been shown to have negligible corrosive ion leaching and permeability to water. However, the cumulative dose rate on the epoxy has to be limited to prevent epoxy degradation. Measurable changes in epoxy liner properties have been observed after a 1 GGy radiation dose. With the general trend of increasing spent fuel pool operating lives and pool licence extensions, this dose limit could be exceeded at some point after the original design life of the pool, and deterioration induced by radiation may be observed. A risk to the epoxy coated concrete walls also exists when the water temperature is maintained higher than 32°C on a regular basis. If epoxy liner deterioration from radiation continues, there is a possibility that water may eventually contact the structural concrete. Thus, a standing programme to monitor the condition of the epoxy coating is necessary. Fuel pools with stainless steel liners likewise require a standing programme to ensure the continued integrity of the liner, with attention on the areas around where the plates are joined by welding to ensure stress corrosion cracking is not occurring. Some of the storage pools with metal liners are provided with leak detection arrangements especially for welded joints.

2.2.3. Double walled containment

Spent fuel pools with double walled stainless steel liners may be employed to preclude the infiltration of contaminated pool water into the groundwater. These pool designs may include a capability for water collection (e.g. a low point sump) in the event that the inner liner is breached. This type of design allows for the early detection and temporary mitigation of fuel pool leaks, while nearly eliminating any potential for release. Identification of the origin of the potential leak may be facilitated through inspection of the interspace area between the two liners. The double walled pool design therefore provides improved protection against spent fuel pool leaks but at an increased cost over that of single-wall facilities, even considering the necessity of instruments for leak chases along weld regions for leak detection and isolation.

2.2.4. Considerations about accessibility for surveillance, maintenance and repair

Inability to access all surfaces of a structure reduces the ability to completely verify physical condition and absence of degradation. Structures that are largely inaccessible (e.g. foundations or walls below grade), represent a primary concern for the use of direct visual inspection as the main inspection tool. In addition, certain degradation mechanisms such as fatigue may manifest and propagate within a structure before any visible signs are displayed. For structures exposed to thermal effects and time-varying or vibratory loads, it may be necessary to supplement visual inspections with non-destructive or destructive testing to examine subsurface conditions. Consideration of the need for visual and supplemental testing has to be given during any facility improvement or modification, to allow incorporation of features that facilitate the necessary accessibility to support fuel, structural and facility surveillance activities.

2.3. STORAGE POOL MAINTENANCE

According to IAEA Safety Standards Series No. SSG-15 (Rev. 1), Storage of Spent Nuclear Fuel [20], the design and operation of a spent fuel storage facility should include an appropriate programme of maintenance, inspection and testing of structures, systems and components (SSCs) important to safety, including physical access to the related spaces. In Sections 2.3.1 and 2.3.2, special consideration is given to items specific to the spent fuel storage pool.

2.3.1. Pool cleaning — filtering and deionization systems

Water treatment is necessary for several reasons:

— To remove suspended solids in order to maintain pool water clarity;
— To remove dissolved radioactive species to minimize the dose to operators;
— To limit buildup of nutrients (phosphates and nitrates) to minimize biological growth;
— To limit buildup of aggressive ions (e.g. chlorides and sulphates) that can initiate corrosion of the stored fuel assemblies or in-pool components.

Water treatment normally includes a mechanical treatment to remove the solid materials contained in the pool, in conjunction with a chemical treatment to extract both radioactive and non-radioactive chemical species dissolved in the pool water (see NP-T-5.2 [18]).

Mechanical treatment of the bulk pool water is generally performed by filters (precoated sand or mechanical), while chemical treatment is realized with ion exchangers (cationic and anionic resin). In some cases, ion exchange is preceded by neutralization. Generally, only single bed organic ion exchange resins are regenerated when saturated. The resultant concentrate may include a boron recycling step, followed typically by evaporation and final encapsulation for disposal. Mixed bed filters and inorganic exchanger resins are also used. In these cases, there is no regeneration phase and the saturated beds are disposed of directly after encapsulation.

The buildup of particulates on the pool floor and walls is removed mechanically by in-pool cleaners. A variety of designs have been deployed from modified leisure pool cleaners. These range from simple suction devices to purpose-built two-stage cleaners with coarse to fine (cyclone) filters. In the case of ion exchange, there are three types of systems in operation:

— Out of pool ion exchange columns;
— Ion exchange floated on top of a precoated mechanical filter;
— In-pool water treatment units.

Examples of the latter include cartridge systems and the French combined ion-exchange–cooling system (so-called Nymphea system).

Special attention is required to avoid the growth of microbiological species that can reduce water clarity or even lead to microbial attack of storage materials. The main factors in limiting biological growth are minimizing the introduction of nutrients (especially phosphates), intensity of lighting in the storage area and temperature. Where bacterial growth does initiate, treatments have varied from the use of biocides to full-scale mechanical cleaning and collection of the biological growth. Operating pools at a high pH also prevents biological growth.

2.3.2. Pool cleaning — sludge removal and vacuuming

The primary sources of the sludge within spent fuel storage basins are degradation of the basin walls, storage racks and the outside of the fuel storage containers, as well as the introduction of dust, sand and silt from the air. Additional sludge volume may be realized by the introduction of particulate matter resulting from backwashing sand filters for the basin water.

This accumulation of non-radioactive material, debris, dirt, sand and silt typically contribute the major portion of the sludge on basin floors. Although the underwater corrosion of the canisters, the storage racks and the fuel cladding itself is not rapid, the steady progression produces contaminants to the basin sludge. Typical spent fuel pool sludge is expected to contain at least trace quantities of activated metals and oxides, fission products, and transuranic elements, in addition to the non-radioactive components of the sludge. In addition, there may be some content of oil and grease from routine maintenance and repair activities associated with fuel handling and transfer and from other operating activities. Insects and rodents may also contribute to the accumulated non-radioactive components of sludge.

The removal of approximately 50 m^3 of sludge from Hanford site's K-Basins was conducted using submersible pumps to move sludge from the basin surfaces to newly constructed submerged steel containers. During the process, operators used tools on long poles and specially designed vacuum heads from a grating above the pool surface. The pumping system included a strainer to ensure material greater than 0.635 cm was not included in the collection containers. In addition, the collection containers were equipped with an inlet distribution manifold, flocculent injection system, and sloped settler tubes on the tank top where water flowed out of the tank in order to minimize carry-over of finer sludge particles back into the basin water.

Major challenges associated with sludge removal from Hanford site's K-Basins included levels of airborne radioactivity that were sufficient to require respiratory protection, loss of visibility from the operating deck (largely overcome by underwater cameras) and difficulty separating debris from sludge. The recovered sludge was treated by heating under moderate pressure to accelerate corrosion of uranium metal to extinction and then grouting the sludge in drums to be disposed as remote-handled transuranic waste [21].

The sludge removal system at Marcoule, France, used a suction head that included water spray nozzles to locally mobilize sludge. The vacuum head was manipulated through a shielded movable plate and rotating sphere device mounted in the basin cover plate. The pump was submerged and the above basin piping was shielded to minimize requirements for protective equipment [21].

At the Idaho National Laboratory (INL), divers performed activities related to the removal of debris and vacuuming of basin sludge with success. The project claims a dramatic decrease in worker exposure and minimal incidence of personnel contamination related to the donning and doffing procedures (corrected). The diver approach resulted in improved schedule management due to the ability of divers to identify and dispose of small and potentially high activity debris without impacting vacuuming operations and project schedule. INL activities included the use of a system employing a submersible pump with a multibag filter unit and integrated flow control instrumentation [22].

2.4. STORAGE POOL INSPECTION AND SURVEILLANCE

2.4.1. Inspection of pool structural integrity

Components such as metallic and non-metallic liners, waterproof materials, joint sealants, and protective coatings and membranes are provided to protect concrete from aggressive fluid and other environmental exposure. These need to be considered in the inspection programme for the pool's structural integrity. In order to ensure that they preserve their function, in addition to the concrete, the storage pool inspection programme should characterize the condition of these components, with maintenance performed as required to preserve their function. For metallic liners, visual inspection should be augmented with ultrasonic measurement of remaining wall thickness and local non-destructive examination (NDE) of any disruptions (e.g. damage) observed. For sealants, coatings and membranes, visual inspection will likely provide the best indication of system performance, condition, adhesion and remaining service life. Should any concerns be identified by the inspection team, vendor representatives for the particular material have to be contacted for input.

Spent fuel storage pools also need to be maintained to avoid leaching of constituents of the pool into the pool bulk water, and to avoid uptake of radionuclides into the pool structure. In this regard, a chemical monitoring programme should consider checks for leaching of concrete or coating or liner constituents into the pool water.

Elements of a storage pool structural integrity programme include:

- Identification of the specific materials and service conditions (environmental and loading conditions) of the structure;
- Identification of codes and standards and natural phenomena hazards or events for the pool design;
- Identification of the service history of the pool;
- Comprehensive degradation evaluation of the materials;
- Periodic inspection to provide a condition assessment;
- Leak detection;
- Repair technologies.

2.4.2. Coating integrity

Concrete spent fuel pools are often coated with an epoxy resin that is designed to provide a physical barrier between the basin water and the underlying concrete. The epoxy lining can mitigate potential degradation of the reinforced concrete basin structure that may occur due to water infiltration. However, when employing an epoxy coating for this purpose, the cumulative dose rate on the epoxy has to be limited to prevent epoxy degradation. Measurable changes in epoxy liner properties have been observed after a 1 GGy radiation dose [23]. With the general trend of increasing spent fuel pool operating lives and pool licence extensions, this dose limit could be exceeded at some point after the original design life of the pool is reached and radiation induced deterioration may be observed. Since 1988, some radiation induced deterioration has been observed at the Pickering NPP in Canada, without any damage on the liner envelope. In that station, a risk of minor damage to the coated concrete walls has been observed when the water temperature is maintained higher than 32°C on a regular basis. If epoxy liner deterioration from radiation continues, water may eventually contact the structural concrete. Thus, a programme to investigate the long term effect of water on concrete has to be implemented.

The condition of the protective coating system used on a liner plate or concrete structure may be verified with adhesion inspection and augmented visual testing methods. Further information on the in-service inspection of protective coatings can be found in Ref. [24]. Also, the use of multiple methods can provide a high degree of damage detectability in coating systems [25].

2.4.3. Stainless steel liner integrity

For stainless steel lining, no corrosion phenomena have ever been observed; however, routine inspection of the liners may be included as part of the in-service inspection. The following are suggestions for evaluation of the integrity of stainless steel liners.

2.4.3.1. Leak testing

The presence of cracks or other through-section disruptions may be identified through leak testing, depending on geometry, if access is available to the back surface of the liner or to the leak testing ducts (behind primary welds). Leak testing may be performed with a variety of techniques, ranging from the sophisticated helium mass spectroscopy to simple soap bubble tests. Leak testing has good detectability of damage when applied locally; however, its use is limited to thru-liner damage. Reference [26] provides an overview of various leak testing methods available.

2.4.3.2. Thickness testing

Ultrasonic waves may be passed into a liner plate or attachments to measure thickness based on the principle that sound passes through certain materials with known velocity. This method is useful for quantifying the remaining thickness of the liner non-destructively. The method has a high degree of detectability to corrosion and planar material loss if suitable procedures are followed and contact surfaces are not heavily damaged. The method described in Ref. [27] may be used.

2.4.3.3. Weld and base metal defect testing

To identify the presence and size of weld defects in the liner plate NDE such as ultrasonic and liquid penetrant methods can be used. Test procedures have been developed by the American Society for Testing and Materials and the American Society of Mechanical Engineers [28]. Damage detectability is a function of size and orientation for each method; continued development of ultrasonics has greatly improved detectability and minimum detectable damage size. Liquid penetrant testing has the least sensitive detectability function.

2.4.4. Aluminium liner integrity

Unlike stainless steel lining, some corrosion phenomena degradation has been detected in aluminium liners in research reactors and fuel storage pools. In most cases reporting degraded aluminium lining, the corrosion process started not in the internal side of the reactor pool but in areas on the external side of the reactor pool where water accumulated, causing corrosion. Several cases of degraded fuel storage pool liners are discussed in Sections 2.4.4.1 to 2.4.4.5.

2.4.4.1. Mexico — TRIGA Mark III reactor

The first documented case of a degraded fuel storage pool liner was in Mexico, at the 1 MW Training, Research, Isotopes General Atomics (TRIGA) Mark III reactor, which has a water filled open pool. Its walls and floor consist of a completely welded aluminium tank embedded in the shielding concrete structure; the aluminium pool liner has a nominal thickness of 6.5 mm. In March 1985, during a routine test, a water leak was found in the primary system of the reactor. The water leak reached a maximum of almost 5 L per hour, then decreased within two months when the holes were blocked by suspended particles that were introduced into the primary system during the leak tests. Hydrostatic testing of the primary cooling system and for the piping of the exposure room, identified the problem in the reactor pool aluminium liner and not the piping system. In order to inspect the walls of the tank, the pool was drained from 7.49 m of water down to a 3.6 m depth, where a localized loss of thickness of the aluminium pool liner was detected. It was determined that the root cause was an involuntary spill of water into the pool, which penetrated into the space between the liner and the concrete structure, enabling an environment for localized corrosion of the aluminium liner from the outside towards the inside of the pool. The section of the liner that was identified as thinning was extracted from the existing aluminium liner. Next, tar and resins were added in the space between the concrete and the aluminium liner. An aluminium plate was then welded into the existing aluminium liner, replacing the thinned area that was removed. Subsequent inspections indicated a loss of thickness in new areas of the liner and signs of localized corrosion starting in other positions. Facility staff evaluated historical data and performed additional inspections, revealing that: (a) there was no significant loss of aluminium thickness in any original or repaired area of the pool liner; and (b) a single point with thickness equal to 2 mm was found (original thickness was 6.5 mm), corresponding to a pore in a padded weld made during the construction of the pool. The maximum corrosion rate, considering the data obtained in 1985 and in 2001, was determined to be less than 0.22 mm/year [29].

2.4.4.2. *Vietnam — TRIGA Mark II reactor*

In 1993, the organization operating the 500 kW TRIGA Mark II research reactor in Dalat, Vietnam, reported corrosion in the aluminium reactor pool tank (on the aluminium pool liner and associated components made of aluminium alloy 6061). An examination revealed that the corroded areas had been caused by mechanical defects made during the replacement of some internal structures in the early 1980s. New visual inspections in 2011 and 2012, using an underwater high resolution video camera system, revealed the following: (a) two of four steel bolts used to tighten the flexible joint of beam port No. 4 were rusted and the results indicated that this rust was developing slowly as a consequence of electrochemical corrosion. However, it was difficult to explain clearly why only two bolts were rusted while the other two bolts still had a shining colour of stainless steel; (b) some pitting corrosion with 2–3 mm diameter on the outside surface of beam port No. 3, probably caused by surface scratches or mechanical defects during the replacement of internal structures in the early 1980s; and (c) pitting corrosion on the pool tank liner, thermal column, beam ports, and in a weld line, where corrosion products of $Fe(OH)_3$ have been observed [30]. As reported to the IAEA Research Reactor Ageing Database (known as RRADB), the operator implemented the following corrective measures to stop the corrosion process: (i) maintaining the water chemistry of the reactor pool and the spent fuel storage pool in good condition by means of a cleaning system, and performing measurements every day during reactor operation and every week during reactor shutdown; and (ii) maintaining visual inspections and cleaning of the reactor pool tank and in-pool components every three months.

2.4.4.3. *Bangladesh — TRIGA Mark II reactor*

Another case of corrosion in aluminium pool structure was reported in the 3 MW TRIGA Mark II research reactor in Dhaka, Bangladesh. In 2009, after 23 years of operation, water was observed flowing inside of the beam tube during the removal of a beam port plug. The cause was identified as corrosion of the internal part of the aluminium tube of the beam port, located about 8 m below the water surface. The corrosion was caused by air and moisture inside the beam tube. The condensate that accumulated in the annular space between the graphite plug and the inner wall of the aluminium beam port created the conditions necessary for corrosion. The stainless steel sleeve used to cover the circumferential gap between the stainless steel and aluminium pipes was not wrapped with sealant during installation. As a result, water vapour from the surrounding concrete found its way to condense into the gap between the graphite plug and the aluminium pipe. The continual presence of moisture transformed the otherwise protective Al_2O_3 layer into $Al(OH)_3$, which cannot prevent air and moisture from seeping through it to attack the fresh aluminium underneath.

The water leak was stopped by putting a rubber strap around the damaged part of the beam port. The corrective actions undertaken by the operator included the installation of an outside aluminium sleeve to protect the outer surface of the aluminium pipe and seal the flow of water from the reactor tank into the beam port. To this end, a split type encirclement clamp was locally designed and fabricated using aluminium alloy 6061, and then installed around the damaged part of the beam port by using a remote handling mechanism. About 48 hours after the installation, the inside of the beam port was inspected with a camera and no trace of water was found. The split type encirclement clamp was designed and fabricated with provision such that it can be dismantled for replacement of the silicone lining and reinstalled. Prior to its installation, several fuel elements had been removed from the part of the core that was in the line of sight with the beam port, in order to minimize radiation streaming from the core. The repair allowed the reactor to return to normal operation safely [31].

2.4.4.4. Indonesia — Kartini reactor

A problem in the 100 kW Kartini research reactor of TRIGA Mark II type in Yogyakarta, Indonesia, which has been in operation since 1979, was discovered in 2001 during an inspection of the aluminium pool liner (visual examination, a hardness survey, dye penetrant testing, and an ultrasonic thickness survey). The inspection revealed two areas of swelling (bulges) observed under the thermal column of the reactor [32]. The thermal column provides a neutron path for experiments that can be conducted in the bulk shielding facility (BSF), an adjacent water pool with a concrete walled structure 2.65 m long, 2.40 m wide, and 3.80 m deep. The BSF has also been used for interim storage of spent fuel. The thermal column consists of a graphite block (dimensions 61 cm × 61 cm × 132 cm), connecting the outer part of the reactor reflector to the BSF.

The defects detected in 2001 were re-examined in more detail in 2004 and 2005, revealing increasing swelling, but examinations performed in 2006 and 2007 showed that the size of the swelling had remained relatively constant since 2005. It was apparent that some element of the reactor block structure was expanding and forcing the pool liner into the reactor pool.

The root cause for swelling was a failure in the BSF which allowed water to flow into several areas in the space between the aluminium liner of the pool and the concrete structure, including the area between the pool and the thermal column. White deposits on the side of the BSF were analysed as sodium carbonate (Na_2CO_3), indicating that water had been seeping through the concrete wall, carrying dissolved material from the concrete that had reacted with CO_2 from the air to form the carbonate. The presence of sodium posed the possibility that the water used to make the concrete structure of the pool contained salt, suggesting the possibility of chlorides in the concrete that could lead to corrosion issues. At that time, no significant physical damage was observed on the internal liner of the reactor pool. Ultrasonic measurements revealed that, despite the swellings, the aluminium liner thickness remained close to the original value and had not suffered any significant corrosion. Remedial actions included drying of the concrete block, removing the humidity that caused the defects and, finally, installing a liner on the BSF to prevent future water penetration into the concrete structure. In addition, it was decided to maintain a programme of periodic inspections of the defects to monitor the rate of progression and take action to repair the affected areas [32].

In early 2009, the BSF has been modified and divided into two parts (wet and dry). The wet part has been provided with an aluminium liner and is used as a water pool for interim spent fuel storage and for radiation shielding of the thermal column. This part faces directly towards the thermal column. The dry part (without aluminium liner) is located behind the wet part. As reported to the IAEA RRADB, the results of periodic inspections performed in 2009 and 2014 showed no water leakage from the thermal column or the wet part of the BSF.

One of the main lessons learned by the organization operating the Kartini research reactor was that periodic examinations of the reactor pool liner are essential to detect issues at an early stage. Degradation processes that occur in the structural elements behind the pool liner can affect the pool liner integrity. It needs to be ensured that the area behind the pool liner remains dry. The BSF attached to the reactor block is a potential source of water ingress to the area behind the pool liner. Any cracks caused by concrete shrinkage or seismic events can provide a path for a water leak.

2.4.4.5. Philippines — PRR-1

NDE such as ultrasonic and liquid penetrant methods can be used to identify the presence and size of weld defects in the liner plate. A leak on the Philippine research reactor PRR-1 at the Philippine Nuclear Research Institute (PNRI) in Quezon City was discovered in 1988. The reactor was originally a 1 MW open pool general purpose General Electric research reactor, which began operating in 1963. In 1985, the reactor was shut down to replace some components to increase the reactor power to 3 MW and convert it to use TRIGA type fuel. The conversion was completed in March 1988, when PRR-1 restarted operation; the conversion was initially found to be successful until several technical problems were identified. The discovery of a major water leak in the reactor pool liner, and other issues that subsequently appeared, led to the extended shutdown of the reactor since April 1988. The liner had been installed after the concrete of the pool structure was poured and was welded on a framework of aluminium I-beams bolted to the pool concrete. Because of accessibility problems with some joints, a section of the liner could not be successfully welded to the thermal column casing, and the joints were packed with epoxy to make them watertight. The analysis showed that, after years of operation, radiation deteriorated the epoxy resins allowing water to flow into the space between the liner and the concrete. The joints were modified to allow their full re-welding, and the liner was repaired. While the joint repair work was being done, corrosion was discovered in other places of the pool liner, requiring construction of a new spent fuel storage pool to allow the original reactor pool to be completely emptied for the necessary repair. Moreover, a seismic fault less than 5 km away from the site became suspected by national authorities to be capable of causing a severe earthquake. In the light of the identification of additional ageing problems and a lack of funding, an internal PNRI study in 2002 recommended stopping the repair work and decommissioning the reactor [33, 34]. All 50 spent fuel elements of PRR-1 were shipped back to the USA in 1999 under the USA FRRSNF Acceptance Program [9].

In 2005, PNRI formally decided to decommission the reactor, and the PRR-1 was accepted as a model reactor for the IAEA's Research Reactor Decommissioning Demonstration Project (known as R^2D^2). The decommissioning plan defined the unrestricted release of the site from regulatory control as the end state of decommissioning [33, 34].

Starting in 2014, however, PNRI developed an alternative option for PRR-1 which will allow use of the existing TRIGA fuel rods in a subcritical assembly for training, education and research. These fuel rods were only slightly irradiated during the conversion testing in 1988 and have been maintained in the facility for more than 30 years [35]. The subcritical assembly which is currently under construction will provide beneficial use for available resources in PNRI with minimized waste generation while preparing for later decommissioning.

2.4.5. Leak detection

Leaking from the spent fuel pool is monitored, either by means of an integrated leak collection system or via the interspace in pools with two walls. In both cases, any recovered pool water may be cleaned up and returned to the main pool [36]. For single-wall or lined fuel pools without an integrated leak collection system, acoustic emission leak testing may be employed to detect leaks. This technique utilizes a piezoelectric transducer to detect the sound produced by the turbulent flow through the leak opening.

For leaks that have escaped detection and have proceeded for longer durations or increased volume, there may be other indications. Monitoring of the volume of makeup water that is required in the fuel pool to maintain a consistent water level may reveal excess loss of basin water. Periodic inspection of accessible areas outside of the basin structure may reveal staining of the concrete structure or stalactite formation indicative of water migration through concrete. Finally, periodic samples from monitoring wells outside the pool footprint and down gradient may contain radioactive material indicating containment failure. Another way of leak monitoring is to compare the loss of water from the pool with the estimated loss due to evaporation and other known losses such as sampling.

2.4.6. In-service inspection

In general, visual inspections and condition surveys for periodic evaluations will view all accessible surfaces of the structure. Reference [37] provides some recommended acceptance criteria for visual inspections of concrete structures. To assist in decision making, it may be necessary to monitor the condition of a structure over a short period of time (less than one year). All significant findings and their classification and treatment need to be reviewed by the engineer in charge in the form of an evaluation report. Structures that are partially lined, protectively coated, or partially or wholly inaccessible need to be carefully evaluated, because several environmental conditions can be present and degradation may be masked or hidden.

The evaluation frequency has to be based on the aggressiveness of environmental conditions and physical conditions of the plant structures. The established frequencies also have to ensure that any degradation related to ageing is detected at an early stage. All safety related structures have to be visually inspected at intervals not to exceed ten years, in accordance with the interval for periodic safety reviews.

A structural ageing assessment report forms part of the activities related to a specific task, namely, structural component assessment/repair technology, in the US Nuclear Regulatory Commission's Structural Ageing Program as executed at the Oak Ridge National Laboratory (known as the ORNL) [38]. It outlines an ageing assessment methodology which utilizes numerical ranking and relative weighing procedures to evaluate, categorize and prioritize concrete structures in NPPs by the importance of their subelemental parts, safety significance, environmental exposure and influence of degradation factor. Use of the methodology was demonstrated through application to three facilities: a pressurized water reactor (PWR) large dry metal containment, a boiling water reactor (BWR) reinforced concrete Mark II containment, and a PWR large dry prestressed concrete containment [38]. Of primary significance to this estimate is the description of the causes of reinforced concrete degradation. The potential causes are chemical attack, efflorescence and leaching, sulphate attack, bases and acids, salt crystallization, alkali–aggregate reactions, moisture changes, freeze–thaw cycling, thermal exposure and cycling, irradiation, abrasion, erosion, cavitation, fatigue, vibration and creep.

2.4.7. Inspection of spent nuclear fuel racks and baskets

Typically, spent fuel assemblies are wet stored within bundles or baskets within the spent fuel storage pool to facilitate bulk handling and movement of fuel within the pool. These bundles and baskets are subsequently stored in storage racks that typically rest on the pool floor. These components need to be designed, fabricated and deployed to preclude preferential attack of the fuel, the racks, the bundles or baskets, and the pool liner. Each of these components is subject to similar corrosion degradation phenomena to that of the fuel assemblies. Therefore, the materials used for construction of these components need to be included in both the corrosion surveillance and the in-service inspection programme, to ensure that localized corrosion attack is averted. Surveillance coupons have to include welded and non-welded specimens and coupled specimens of similar and dissimilar materials in order to replicate potential configurations present within the basin. A comprehensive in-service inspection of SNF racks and baskets has to be conducted periodically to verify the results of the corrosion assessments that are based on the corrosion surveillance coupons. The interval between inspections should not exceed ten years and may be shortened based on results of the corrosion surveillance and water chemistry monitoring programmes. Guidance on the fuel handling processes can be found in IAEA Safety Standards Series No. NS-G-4.3, Core Management and Fuel Handling for Research Reactors [39].

2.4.8. Inaccessible structures

Partial or totally inaccessible structures require that a different approach and emphasis be taken towards in-service inspection. The most obvious shift in emphasis is away from visual inspection towards environmental quantification. A possible inspection approach for a partially accessible structure could consist of the following steps:

(1) Perform visual inspection on accessible surfaces;
(2) Perform NDE on accessible surfaces, as directed by visual results;
(3) Study original design and identify key durability features;
(4) Quantify environmental conditions potentially affecting inaccessible portions of the structure;
(5) Perform initial review of data to identify any shortcomings;
(6) Conduct further inspection and testing and evaluate the structure's condition using all available data.

When the structure is partially accessible and site investigations have found the environmental exposure to be non-aggressive, further action is probably not required. Partial inspection (and knowledge) may provide enough information to gain an understanding of structural integrity. However, if site investigations determine that the environment is potentially aggressive and the structure is inaccessible, it may be necessary to expose a portion of the structure via boring, soil and structure removal or other effort to evaluate the extent of degradation. In a pressure-retaining boundary, coring through the liner to obtain samples for testing is discouraged, as this can compromise the leak-tight integrity of the boundary. However, this may be the only means available to investigate concrete condition.

Surfaces of structures that are completely lined with a metal liner for fluid retention, and which are not part of a pressure boundary, may be locally assessed via core sampling through the liner. As the liner is generally not considered to be a structural element in design, local liner repair efforts after coring should not be substantial. This form of core sampling could prove quite valuable as the liner condition, liner–concrete bond, and concrete conditions below the liner (and possibly that of embedded reinforcing steel) are all determinable with a single coring operation.

2.5. FACILITY LIFETIME EXTENSION

Lifetime extension of a nuclear facility requires a comprehensive assessment of the current state of the facility with a demonstrable prediction of its future behaviour and its structural components. An extensive review of facility plans and as-built drawings are required to identify construction materials and performance characteristics of components that are pertinent to structural integrity of the system and material confinement. Information required includes design loads and load bearing capability of the facility. Loads on the operating floor may include live loads, equipment loads, and supports for the part of the overhead monorail system not suspended from the roof. Considering that spent fuel is typically stored in racks that bear directly on the basin floor, identification of load bearing members for these loads is necessary.

After assessing the load requirements of the facility, a detailed structural analysis is required to demonstrate continued structural integrity with consideration of current seismic codes and requirements, as well as wind and tornado analyses. The seismic and wind analyses require the input mentioned previously relating to construction materials and design characteristics.

In addition, considerable effort and tests may be required to demonstrate the current mechanical and structural condition of the structural members of the fuel storage facility. This includes a detailed visual inspection of the facility, NDE testing, destructive analysis including selective sample extraction and mechanical and chemical testing, and the employment of subject matter experts to make engineering judgements where analytical data are not available and cannot be obtained.

The most effective approach to facility lifetime extension includes the maintenance of detailed as-built drawings and plans for the facility, including procurement specification and acceptance criteria, and the use of a regular and periodic facility inspection programme. The inspection programme has to be augmented with materials sampling (core harvesting) to evaluate potential degradation phenomena of calcium leaching (water side) and carbonation and sulphate attack (air and soil side) to evaluate the depth of degradation of the concrete. Results of the depth of degradation have to be considered in revised structural analyses, as appropriate.

2.5.1. Recoating

Depending on the characteristics and conditions of the spent fuel storage pool, recoating may be an option for lifetime extension. The only information available in the literature about recoating spent fuel storage pools is the result of a series of tests run at INL [40]. These tests were conducted to evaluate 14 candidates for underwater recoating of the INL spent fuel storage pools as part of a deactivation programme of specific fuel basins. The candidates were evaluated on ease of application and on the adherence of the applied coating to the concrete substrates. It is important to note that the main purpose for recoating the storage pools at INL was to minimize the risk of airborne contamination after emptying the pools by trapping residual contamination on the basin walls. These tests did, however, demonstrate the overall feasibility of underwater recoating. A more detailed description of the tests and analysis of their results can be found in Ref. [40].

2.5.2. Leak repair

2.5.2.1. Metal liners

Stainless steel and aluminium are the two primary liner materials for pools. Stainless steel liners are expected to be nearly immune from degradation that might lead to a leak. The one exception would be a condition where weld joints in the liner were excessively heated during fabrication, resulting in a sensitized condition and rendering it susceptible to cracking. Typically, thin-walled stainless steel weldments are not susceptible to stress corrosion cracking.

Aluminium lining is subject to corrosion attack if the water quality in the pool is poor. In addition, corrosion attack of the aluminium lining may occur from outside the pool inwards if water accumulates in the interface between the lining and the concrete structure (see specific examples in Section 2.4.4). For this reason, it is important to maintain water quality in these systems even during periods with no spent fuel with aluminium cladding in storage. Additionally, it is recommended that an inspection and maintenance programme be developed and maintained for the pool lining.

2.5.2.2. Leaks through concrete

Leaks have occurred in both power reactor and research reactor concrete fuel storage pools [19]. Through-wall cracks from the water side to the exterior can occur from shrinking and settlement. These cracks typically do not adversely impact the structural integrity of the pool. Minor cracking could lead to stagnant water contacting the reinforcement bars in the steel but, as the water tends to reach a high pH, it would not be aggressive enough to cause corrosion of the carbon steel reinforcement bars. Through-wall cracking that would result in water flow past the bars may cause corrosion that, over time, would result in an unacceptable loss of structural integrity. Brown staining of water that has leaked from the pool is an indicator of reinforcement bar corrosion.

Leaks can be detected by various methods including visuals or sensors. Effects of the leaking pool water need to be evaluated in terms of water replacement capability, corrosion of reinforcement bars and radiological release.

Two primary types of repair are viable. Localized patching of cracks with grouting or other patching compound to eliminate the leak is a simple fix. For a more robust, certain repair, the installation of a pool liner is recommended to avoid the nuisance of leaking cracks. A structural evaluation has to be performed to ensure the net section and any degradation of the reinforced concrete is acceptable as part of the leak repair. In the case where the structural integrity of the degraded concrete is severely compromised, a structural repair is imperative.

3. TRANSITION FROM WET TO DRY STORAGE

There are several considerations in assessing options of transferring SNF from wet storage to dry storage, such as the condition of the fuel, the basin and the facility. Additionally, there are political and policy issues that will also provide some bearing on this decision. Typically, a decision is made to move to dry storage when the pool has reached maximum storage capacity and after reracking and fuel consolidation options have been exhausted. Other circumstances that necessitate dry storage are when the fuel pool is being closed or decommissioned, based on either a political or policy decision or fiscal considerations related to maintenance of the pool or the facility. At the point when a decision is made to no longer actively maintain the water quality within the basin, it is recommended to transfer the fuel to dry storage.

Fiscal considerations include costs associated with resin regeneration and replacement, maintenance of an effective facility 'in-service inspection' programme, a water chemistry control programme, a fuel surveillance and inspection programme, physical maintenance, upkeep and repair of the storage facility, including the fuel pool. In addition, there are fiscal considerations related to the integration of the storage canister design into the disposal strategy. Moving to dry storage in canisters that may not qualify as a disposal option, and therefore necessitate repackaging, may preclude dry storage prior to the development of acceptance criteria for spent fuel canisters within that disposal option.

The heat generating capacity of the fuel through radioactive decay requires consideration. Although this capacity decreases over time (see Fig. 2) [41], the heat removal capability within a dry storage system is significantly lower than that of wet storage due to the reduced heat transfer capability of the backfill gas and air as compared to bulk water. The storage of fuel in large water filled pools provides a mechanism by which the decay heat load is passively cooled through convective flow within the pool. Thus, dry storage can lead to higher fuel temperatures. The peak temperature of fuel during storage is an important consideration to preclude the possibility of partial or complete melting of low melting point fuel cladding.

The final consideration to be made relates to the condition of the fuel. There are several strategies to manage damaged fuel without the necessity of dry storage. Fuel with minor breaches, and with arrested degradation activity, may be successfully stored for extended times in spent fuel pools while the pool water chemistry is actively monitored and controlled. Fuel with significant degradation or damage may be stored in isolation canisters within the spent fuel pool. This option protects the pool chemistry by isolating the degraded fuel. Work is ongoing at SRS to develop a fundamental understanding of the degradation mechanisms and rate of degradation that occurs with damaged fuel that is being stored in isolation canisters. Any degraded fuel that exhibits aggressive degradation, even in good quality water, is a strong candidate for removal from wet storage to be placed in a dry, inert atmosphere consistent with dry storage. These conditions may occur in fuel with poorly brazed joints or fuel that is physically defective (i.e. inferior material or design features such that galvanic or other localized attack cannot be mitigated in good quality water).

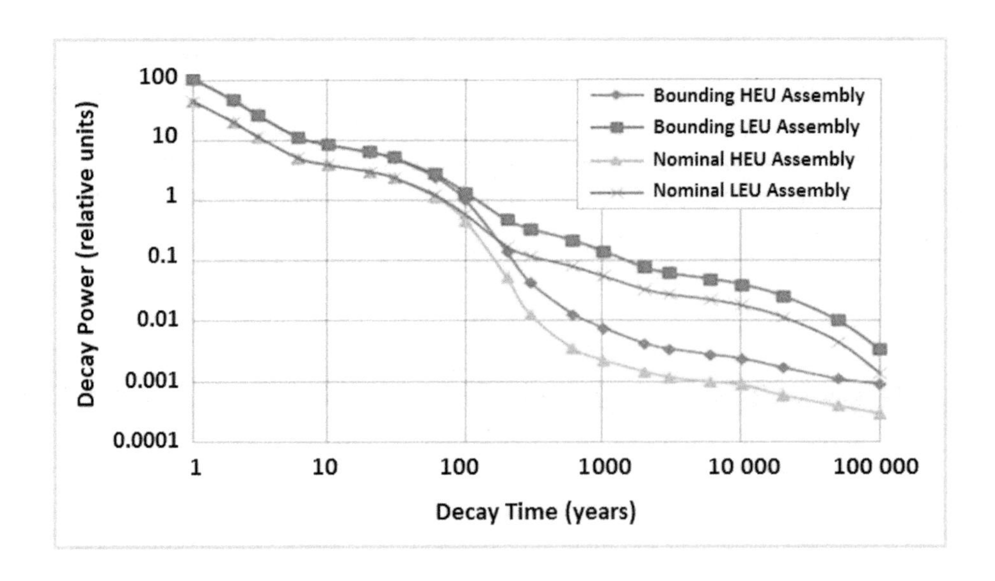

FIG. 2. Decay heat production as a function of decay time for SNF [42].

3.1. PREDRYING CONDITIONING

During wet storage in the pool, the surfaces of the spent fuel assemblies may have deposits consisting of sludge, debris and corrosion particulates. These sludge and debris, if not removed, are detrimental to the subsequent dry storage. When retrieved from the storage pool, the SNF assemblies need to be separated from these particulates and debris before being loaded to the storage container.

For fuel retrieved from pools with poorer water quality, a rinse step in high quality water (i.e. low content of chloride and sulphate ions, neutral pH and low (<50 μS/cm) electrical conductivity) is recommended to reduce potential for pitting corrosion [43]. Any organic materials has to be removed to avoid potential issues associated with radiation decomposition of organic materials.

More thorough cleaning of the retrieved fuel assemblies can be accomplished by using a cleaning system similar to the primary cleaning machine used at Hanford site, USA [44]. The primary cleaning machine provides both mechanical agitation and flushing by slowly tumbling the fuel canisters past high pressure water jets. The cleaning actions dislodge corrosion products, corroded metal pieces and loosely adhered materials from the surface of the fuel elements that are wet stored.

3.2. DRYING OF SPENT NUCLEAR FUEL FOR TRANSITIONING TO DRY STORAGE

Transfer of RRSNF from wet storage to dry storage requires removal of the residual water during the handling and repackaging process. This is the most important step in transferring the spent fuel from wet to dry storage. The objectives of the drying processes used are [42]:

- — To preclude geometric reconfiguration of the packaged fuel;
- — To prevent internal damage to the canister from overpressurization or corrosion;
- — To minimize hydrogen generation or materials corrosion that could be a problem during transport or repository handling operations.

3.2.1. Water forms

The residual water that may exist in a spent fuel container can be in the form of free water, physisorbed water and chemisorbed water. Free water (i.e. not chemically or physically bound) is more easily removed from the fuel than other forms [45].

Physically absorbed water is the water that is physically bound by capillary forces or other weak forces to internal or external surfaces of solid material. Physically absorbed water is found on all surfaces of the spent fuel and the container. RRSNF cladding with cracks, surface corrosion, porosity, or oxide spallation may hold significant amounts of physisorbed water. Typical water concentrations are about 0.03 to 0.05 g/m^2 per monolayer. In addition, water may be held between parallel fuel plates by capillary forces. Assemblies with small interplate gaps can have significant water hold-up that may prove challenging to remove during the drying process. However, the binding forces for physisorbed water are relatively weak, and the water layer, as well as the capillary water, can be removed at relatively low temperatures with adequate vacuum drying or forced-gas drying techniques [42].

Chemisorbed water may exist as a hydroxide or hydrate in the native oxides or corrosion products on the fuel, cladding or container materials. The dehydration of hydroxides occurs via the reformation of water molecules, which are released from the lattice at temperatures or ionizing radiation corresponding to the specific bonding energy of the compound. For RRSNF with aluminium cladding, the surface hydroxides include gibbsite ($Al(OH)_3$ or $Al_2O_3•3H_2O$), which begin dehydrating near 100°C, and boehmite ($AlOOH$ or $Al_2O_3•H_2O$), which is stable up to about 340°C [42]. Complete removal of chemisorbed water generally requires much higher temperatures than are achieved during typical drying processes.

Removal of bound water will only occur when the specific threshold energy is applied to break the bonds involved and release the water. For SNF, this threshold energy may come from the combination of thermal input and ionizing radiation. RRSNF generally has relatively low burnup and low decay heat. Dry storage temperatures and radiation levels may be so low that water radiolysis and secondary oxidation reactions may not occur or may be insignificant [42].

3.2.2. Drying technologies

The methods that may be employed to remove residual water from spent fuel retrieved from pool storage include:

— Drip drying;
— Heating;
— Air drying (hot and cold);
— Forced-gas dehydration;
— Vacuum drying (hot and cold).

The first three drying methods listed above are only capable of partial removal of free water in the canister to varying extents. As such, they are considered only for vented dry storage systems. For sealed dry storage designs, forced gas dehydration or vacuum drying techniques are used.

3.2.2.1. Drip drying

The spent fuel assemblies are often placed into a basket or storage container underwater before being lifted from the pool. Drip drying refers to the process of draining and evaporating of residual water when the loaded basket is left in air for a period of time. Evaporation of water is promoted by the decay heat of the spent fuel. Although drip drying is the simplest drying method requiring no additional drying equipment, it has limited effectiveness. Drip drying is ineffective to remove free water held in the dead-end locations of the fuel assembly and the basket, as well as the free water contained in porous corrosion products or inside fuel with breached cladding. Therefore, drip drying is often used as a preparatory step to remove the bulk of free water for subsequent applications of other drying methods, which are discussed in Sections 3.2.2.2 to 3.2.2.5.

3.2.2.2. *Heating drying*

To improve the drying effectiveness, heating can be applied during the drying process. In this drying process, the fuel basket is typically placed inside a metal cylinder in the drying station. Electric heaters installed on the external surface of the metal cylinder heat up the fuel and basket to accelerate water vaporization. Heating, although an improvement over the drip drying method, is of limited application because:

— Heat transfer from the heaters to the fuel occurs primarily by conduction. The multiple air gaps between the heaters and the fuel make the heat transfer less effective and uneven, and may cause localized overheating of the fuel to temperatures exceeding the temperature limit as well as thermal stress limit for fuel assemblies.
— Heating may not be suitable for damaged fuel with reactive fuel material (e.g. uranium metal). At elevated temperatures, the exposed fuel will react with air at an accelerated rate causing additional damage to the fuel and the cladding with concomitant release of fission products and oxidation of the fissionable material.

3.2.2.3. *Air drying*

With the air drying method, dry (typically hot) air is introduced into the storage container and in direct contact with the spent fuel. Introduction of convective heat transfer significantly enhances the drying efficiency and effectiveness. The relative humidity of the air inlet and the exhausted gas is monitored during the drying process. When the relative humidity of the exhaust gas stabilizes at a value close to that of the inlet air, the drying process is terminated.

3.2.2.4. *Forced gas dehydration*

Forced gas dehydration utilizes recirculated non-reactive gas to dehydrate the loaded spent fuel canister. This drying technique utilizes a superheated non-reactive gas (e.g. nitrogen or carbon dioxide) or one of the inert gases (including helium, argon, krypton, or xenon) that is introduced into the canister cavity, where it picks up moisture from the water remaining in the canister. The wet non-reactive gas is circulated through a condenser module to extract the majority of the water vapour from the gas, followed by a demoisturizer module in which the gas is cooled to a prescribed temperature, T_C, in order to freeze dry the gas, removing remaining water to a specified limit. The dry gas leaving the demoisturizer is superheated to a temperature, T_H, prior to re-introduction into the canister cavity. In addition, the added energy introduced by the superheated non-reactive gas will further instigate the evaporation of liquid water within the canister cavity.

A temperature of −6°C corresponds to a water vapour pressure of 400 Pa, which is used by the US Nuclear Regulatory Commission in evaluating the drying and dryness levels for dry storage packages prior to backfilling with inert gas [46]. Therefore, a T_C value of −6°C, set as the low temperature target for gas exiting the demoisturizer, provides sufficient dryness of the non-reactive gas being circulated through the canister as to affect safe dry storage of SNF canisters, since the vapour pressure within the canister will tend towards the 400 Pa value as the system is used. The high temperature, T_H, could be set as high as 149°C. Higher T_H values result in increased drying efficiency of the system and therefore decreased drying times. A detailed thermal analysis needs to be performed in order to optimize T_H, with consideration of maximum fuel temperature, canister pressure, and other potential limitations. The benefit of this drying technique is that the use of forced non-reactive gas in the drying process precludes the necessity of adding heat to the system to avoid water freezing during the drying process (beyond the heat required to superheat the non-reactive gas prior to introduction into the canister cavity). The high and low temperatures and system pressure during the drying process can be dynamically adjusted to optimize drying results while assuring operation safety.

3.2.2.5. *Vacuum drying*

In vacuum drying, the gas pressure is reduced below the vapour pressure of the water to evaporate the liquid phase. As the vacuum pump turns on, the pressure inside the spent fuel container drops rapidly and then remains constant at an equilibrium value as the water inside the container starts boiling. Another sharp drop of the pressure occurs when the contents become dry and no more water can evaporate from within the canister. When a final pressure of about 133.3 Pa is reached, the vacuum system is isolated. The container pressure would rise if water trapped beneath spent fuel cladding discontinuities were subsequently released. If the vacuum can be maintained after a certain period of holding time, the container is considered dry.

During vacuum drying, the liquid water may undergo a considerable temperature drop as it evaporates from its liquid phase; this can cause the water to freeze, especially for spent fuel with lower decay heat. Drying operation design steps may be necessary to prevent the water from freezing in the container or in the vacuum lines. Drying procedures with thermal homogenization steps, such as helium gas backfill or the use of other hot inert gases, usually prevent ice formation. Throttling of vacuum pumps to slow the rate of vacuum drying also helps to prevent ice formation. RRSNF generally has relatively low burnup and low decay heat, and the use of a specialized heated drying process is required to prevent ice formation.

Below are two examples of vacuum drying processes for RRSNF.

Example 1: Drying fuel at the United States Department of Energy (USDOE) Hanford site

The following method is being used at the USDOE Hanford site to dry Zircaloy and aluminium alloy clad, metallic uranium fuel elements for interim dry storage [47]:

- The retrieved fuel assemblies are first cleaned inside a 'tumbling washer' and rinsed with high pressure spray to remove exterior crud and corrosion products.
- Cleaned spent fuel assemblies are loaded underwater into baskets and placed into a multicanister overpack (MCO) for subsequent drying and storage. The MCO is closed with a screwed shield plug, and is received at the drying facility in a sealed transport cask.
- The MCO within the cask is vented to the cask headspace, which was purged and filled with 20.68 kPa (~3 psi) helium at the storage basin after loading of the MCO was completed.
- Warm water is circulated through the annulus between the MCO and the liner of the cask to prevent freezing any water within the MCO during vacuum drying. The water enters underneath the MCO base, flows upward through the annulus and exits at the top. Average water temperature is maintained at ~45°C.

The major process steps at the drying facility are as follows [42]:

(1) The cask headspace is vented through a venting system and purged with helium, after which the cask lid is removed.
(2) The vacuum drying system is connected to the process ports in the MCO shield plug and the tempered water system is connected to the cask.
(3) The tempered water system heats the cask and the MCO to ~45°C by circulating warm water through the annulus region between the MCO and the cask surrounding it.
(4) When the MCO temperature reaches 45°C, the bulk water in the MCO is removed and transferred to a receiving and treatment system.
(5) Following removal of the bulk water, the MCO is purged with helium and the vacuum pump is started. A helium purge of ~0.71 L/s (1.5 scfm) is maintained until the pressure in the MCO drops to below 1000 Pa (7.5 torr), after which it is secured while the vacuum pump continues to run.

(6) Upon reaching a pressure below 66.66 Pa, the MCO is isolated and a pressure rebound test is performed. The MCO pressure must remain below 306.64 Pa for 1 hour or the drying cycle (step 5) is repeated.
(7) If the rebound test is satisfactory, the MCO is backfilled with helium above atmospheric pressure. The cask or MCO is allowed to cool down using the tempered water at ~15°C. A final helium backfill to a set pressure of 77.57 kPa is performed. The MCO is then sealed and moved to the interim storage location.
(8) Failure to meet this requirement results in repeating the drying cycle from step 5 to 7.

Example 2: Drying fuel prior to transfer to HLW treatment and storage building, Netherlands

The following method is used at the research reactors in Petten (High Flux Reactor) and Delft (Hoger Onderwijs Reactor) in the Netherlands for vacuum drying the RRSNF inside the CASTOR MTR2 metal cask, prior to transfer to the HLW treatment and storage building (HABOG) [48]:

(1) Start the vacuum pump and the recorder to record the pressure;
(2) Check progress of drying on the recorder;
(3) When the pressure reaches the value of 100 Pa, continue drying for one hour;
(4) Close the valves of the vacuum pump and write down the value of the pressure indicator as P_1;
(5) After 15 minutes, write down the value of the pressure as P_2 — calculate $\Delta P = P_2 - P_1$;
(6) If $\Delta P > 20$ Pa in 15 minutes, start the vacuum drying again by opening the valves;
(7) If $\Delta P < 20$ Pa in 15 minutes, continue with the vacuum drying for 1 hour to ensure that the dryness requirement is fulfilled;
(8) Close the valves and write down the pressure value, P_1, and after 15 minutes, the value P_2 — again, the ΔP should be less then < 20 Pa in 15 minutes to prove that the dryness requirement is fulfilled and if it is not, start the complete procedure over again.

The above process is performed without the need for adding external heat to the drying process, as the decay heat from the fuel and filters is sufficient to preclude freezing during drying.

3.2.3. Dryness verification

Due to the temperature limit (200°C) allowed for the aluminium clad RRSNF during fuel handling and storage, high temperature (>340°C) drying to remove chemisorbed water is not practical. For a sealed dry storage system containing RRSNF, the practical dryness requirement is to remove the free water and most of the physisorbed water from a sealed storage container.[2]

For a typical forced-gas dehydration process, dryness is indicated by the relative humidity or water vapour pressure measured from the canister exhaust gas as it exits the canister during the drying process.

For a standard vacuum drying process, adequate water removal is normally evaluated using vacuum pressure rebound measurements, similar to the two examples described in Section 3.2.2.5. The pressure rebound measurements consist of showing that an evacuated container loaded with spent fuel will retain vacuum for a specified period without a pressure rise greater than a specified limit. Container size and fuel characteristics need to be considered in specifying the test pressure, hold time, pressure rise and repetition. In addition, water vapour pressure may be measured in-line at the exhaust during drying and may provide data on how much water remains during the drying process.

[2] Although the release of the chemisorbed water during long term dry storage of RRSNF is expected to be insignificant due to the low decay heat and radiation field of RRSNF, it may need to be confirmed on a case-by-case basis for the specific fuel to be stored.

The following supplementary measurements may also be included, when necessary and available, for dryness verification during storage [42]:

— Container pressurization data over the storage term;
— Sample analysis of the internal gas composition of the container or cask over the storage term, and determination of the hydrogen concentration (as an indicator of the water that was present and released as a result of radiolysis or chemical corrosion within the container).

3.3. INERT GAS CONSIDERATIONS

Another important issue in transferring the spent fuel from wet to dry conditions is the selection of the inert gas used for backfilling the storage canister or simply to fill the environment around the fuel in vault type storage. The inert gas (helium, argon, nitrogen, or some mixture of these) used for leak detection has to be moisture-free and a commercial grade of at least 99.99% purity. This level of purity is essential to limit the degradation of the aluminium cladding during its storage lifetime. The canister may be filled to pressures at or above atmospheric pressure. Higher pressures may be required for early detection of canister leaks.

4. DRY STORAGE OF RESEARCH REACTOR SPENT FUEL

Just as the selection, design and operation of a research reactor is based on the intended mission of the reactor, the spent fuel dry storage facility design and operation is based on the specific needs of the fuel and the intended role of dry storage in the overall spent fuel management strategy. Considerations discussed in this section include technical design, safety and licensing, and economics.

4.1. DRY STORAGE TECHNOLOGY

There is no one design of a dry storage facility suitable for all research reactor fuel types. Some countries — like Finland, Norway, Italy and Canada — have traditionally used dry storage vaults to store RRSNF, following cooling in wet storage conditions for a few years [49–52]. Dry storage vaults were also used in Australia during the operation of the High Flux Australian Reactor [53].

The concept of dry storage using casks — a sealed metal cylinder containing several units of RRSNF which also serves as a radiological shield — has been selected by some organizations operating a research reactor for storage of RRSNF. In some countries, like Brazil and Chile, the decision was taken to use a dual purpose cask (for transport and storage), with the intention that it could be used, at a later time, to transfer the spent fuel either to the disposal site or to a reprocessing facility [8]. In Germany, the decision to dry store RRSNF was taken in the 1970s, when a reprocessing facility was planned in the country. Passive safety and aircraft impact resistance features were required for the spent fuel reception area. This led to the idea of storing spent fuel elements in casks that fulfil the safety functions for storage and transport. Storage of these casks would take place in simple but solid buildings naturally cooled by a self-regulating air flow instead of using a pond with a forced water circulation [54].

The Netherlands, following the decision to keep RRSNF in storage for at least 100 years before making a final decision on its disposal, selected to use only storage casks, piled within metallic silos, in an arrangement that uses a passive cooling system to remove the residual heat produced by the spent fuel [55].

A different approach to dry storage has been applied in Hungary, where the SNF was temporarily stored at the Budapest Research Reactor (BRR). As the storage pool had a limited capacity, another storage pool had to be constructed. At that time, a decision was made by the reactor management to change the storage mode from wet to semidry, to extend the safe storage period of the RRSNF. This semidry technology used a tube-type capsule made of an aluminium alloy which could accommodate one EK-10 Russian type fuel element, one triple Vodo-Vodjanoj Reaktor (VVR) type assembly, or three single VVR type assemblies. After undergoing a powerful drying procedure (heated by eddy current), the capsule was backfilled with dry nitrogen gas, closed with a shrink-fit capsule head, welded and then returned to the wet storage pool. According to Hungarian experts, the adoption of the semidry storage methodology allowed for the safe extension of the storage of the RRSNF for at least another 50 years [3].

Experience has shown that the technology required to avoid any degradation of the RRSNF while in storage in a dry environment for long periods, is expensive and complex. Consequently, dry storage may not be a feasible option for countries with very small nuclear programmes, especially ones with only one or two low power research reactors.

Dry storage using casks presents many advantages compared to other options. However, it requires some infrastructure that may be expensive, especially when the spent fuel assembly discharge rate from the reactor is not very high (i.e. below ten fuel elements per year). Corrosion damage, gas pressurization, metal embrittlement and radiation damage are mechanisms that have the potential to eventually degrade cask performance. Therefore, it is necessary to demonstrate that the degradation suffered over the storage period (100 years or more, in accordance with the RRSNF management programme) does not adversely affect the cask's ability to remain leak-tight during storage and eventual transport of the SNF. Also, to avoid any harmful developments, it is necessary that the RRSNF be properly dried and the cask be backfilled with an inert gas, sealed and maintained under controlled storage conditions. Sometimes a hot cell is used for the loading and drying operations, and attention has to be paid to the security of the RRSNF after the reactor reaches its end of operating lifetime.

4.1.1. Dry storage options and technologies

As opposed to wet storage, dry storage systems contain the spent fuel in a dry and often inert environment to limit oxidation of the fuel while in storage and dissipate the decay heat of the stored spent fuel passively through a combination of radiation, conduction and natural convection.

The design options currently in use for dry storage of spent fuel from research reactors can be divided into four types: modular vault, drywell, metal cask and concrete cask. These options are presented in Sections 4.1.1.1 to 4.1.1.4.

4.1.1.1. Modular vault

This is an above ground, large ventilated concrete structure. An example of this specific type of dry storage is the HABOG storage facility[3] built in the Netherlands, and shown schematically in Figs 3 and 4.

As shown in Figs 3 and 4, the spent fuel is stored in the vaults designated for wastes that generate heat. The vault storage system consists of an array of stainless steel wells filled with stainless steel canisters that contain storage baskets. Each storage basket holds up to 33 SNF assemblies. After vacuum drying, the basket and the canister are weld sealed and backfilled with helium gas. The storage well is filled with argon gas. The decay heat of the stored spent fuel dissipates to the surface of the storage well and is then passively removed by natural air convection. The cooling air enters through air inlets that lead to the bottom of the storage wells. As the air heats up, it rises by natural buoyancy and exits through the ventilation shaft at the top. Radiation shielding is provided by 1.7 m thick, concrete walls reinforced with

[3] HABOG is a vault facility for storage of canisters, rather than for storage of bare fuel elements.

steel and the shield plugs at the top of the storage wells. The facility was designed for safe dry storage of the spent fuel for a period of at least 100 years [54, 55].

The vault dry storage design is a well established and proven technology that has been widely used for dry storage of spent fuel from commercial power reactors since the 1970s in Canada, France, Hungary, the United Kingdom and the USA.

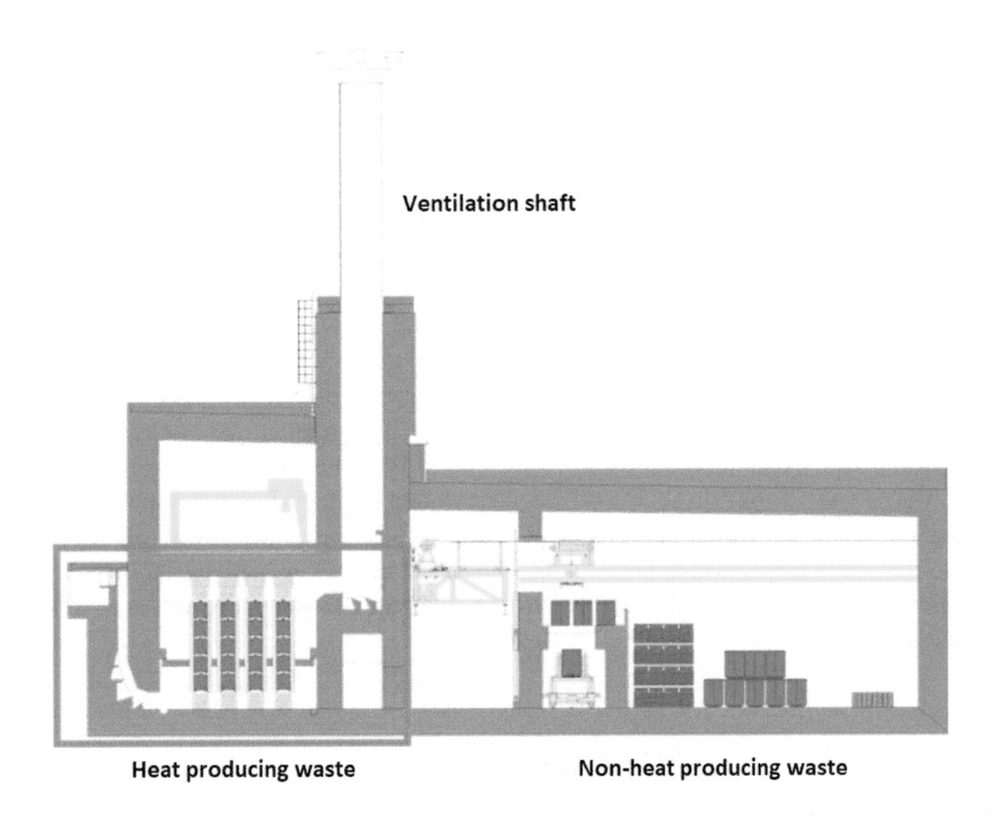

FIG. 3. The HABOG dry storage facility in the Netherlands (courtesy of COVRA N.V.).

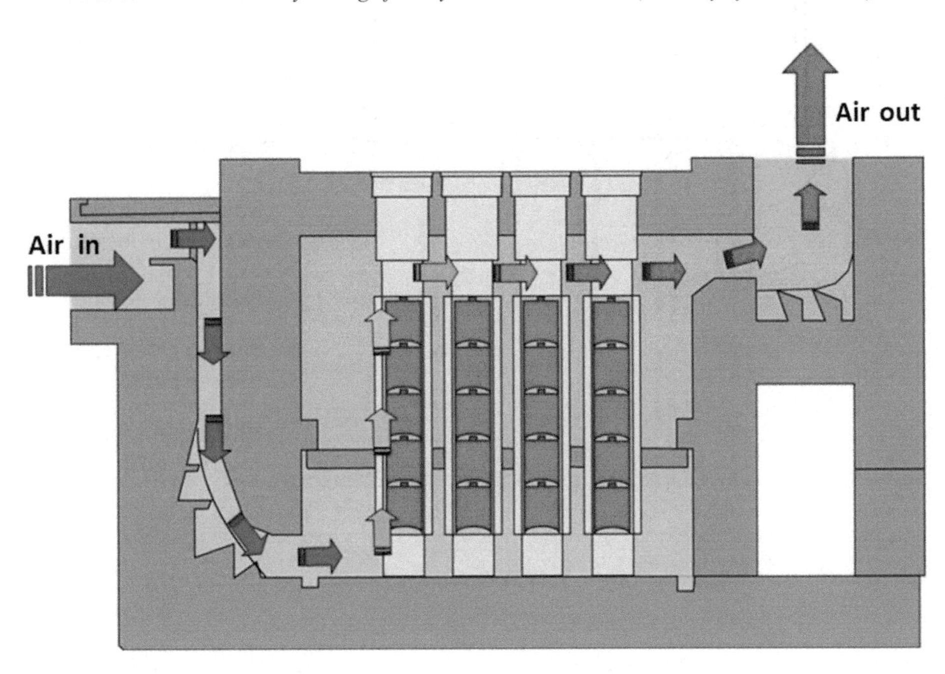

FIG. 4. The passive cooling system in the HABOG dry storage facility (courtesy of COVRA N.V.).

4.1.1.2. *Drywell*

In drywell storage, the spent fuel is placed in individual canisters or cans that are then placed in storage pipes. The storage pipes, also called drywells, are buried in the ground. The ground provides the required radiation shielding and serves as a heat sink for the spent fuel. The drywell storage technology has been successfully used for dry storage of RRSNF in Australia, Canada, Japan, Norway, and the USA.

A schematic of the Australian Nuclear Science and Technology Organisation (ANSTO) drywell storage facility is shown in Fig. 5 and serves as a good example of drywell storage of RRSNF. Built in 1968, ANSTO's dry storage facility, which is used for the interim storage of the RRSNF from the High Flux Australian Reactor, is one of the earliest drywell storage facilities for RRSNF in the world. All the spent fuel assemblies stored in this facility were retrieved for repatriation by 2007. Inspections of the retrieved spent fuel prior to transport demonstrated that the fuel assemblies had been maintained in sound condition with negligible deterioration after about 40 years of dry storage [53].

The ANSTO facility consists of an insulated structure that protects the drywells. The 50 drywells are arranged in a square lattice and are flush with the ground. The wells were bored into the sandstone bedrock. The boreholes are lined with stainless steel tubes, which are sealed at the bottom and closed at the top by a shielding plug bolted to a mild steel collar set in concrete. Two spent fuel elements were placed into an open ended stainless steel can with a hole to allow cooling gas to circulate. Eleven cans were then stacked one above the other within a liner tube. This liner was vacuum dried and backfilled with dry nitrogen. The nitrogen in each well was vented and sampled for ^{85}Kr annually; the well was then re-dried and backfilled with a new charge of nitrogen [56]. Decay heat was passively removed by buoyancy flow of nitrogen gas within the liner, and by radiation and conduction heat transfer to the liner tube. Radiation shielding was provided by the surrounding bedrock which also serves as a heat sink for the liner tubes.

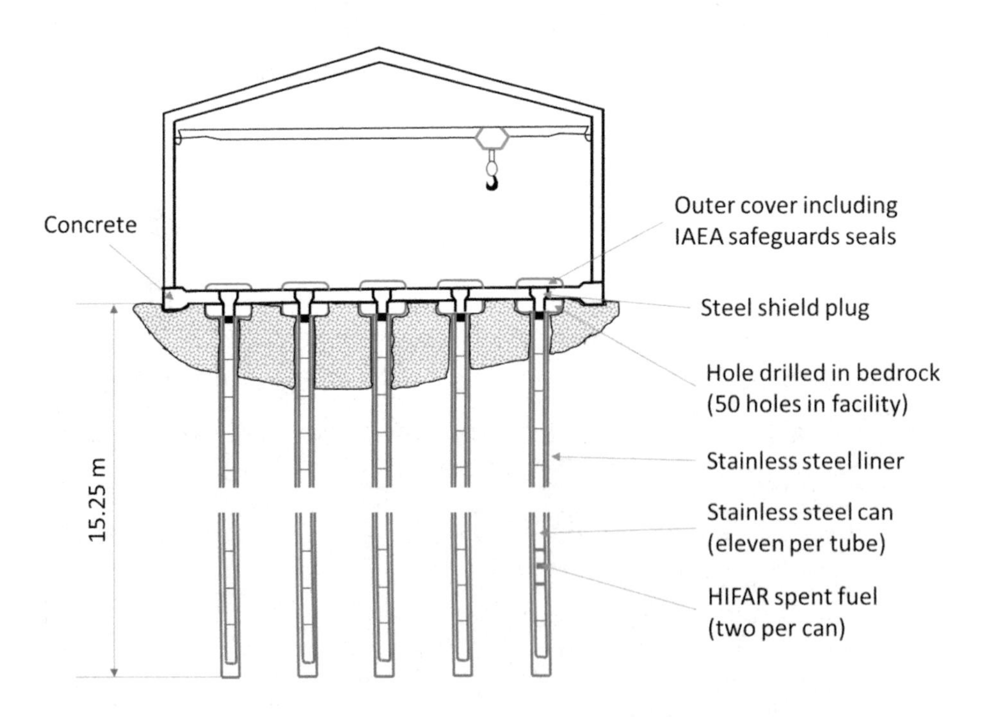

FIG. 5. Schematic of ANSTO's dry storage facility (courtesy of ANSTO).

4.1.1.3. *Metal cask*

In the metal cask design, spent fuel is surrounded by inert gas inside a metal cylinder that is either welded or bolted closed. Each cylinder is surrounded by additional steel, lead and other material to provide radiation shielding to workers and the public. The decay heat of the spent fuel is passively removed to the environment primarily through a combination of radiation and conduction heat transfer mechanisms. Metal casks, CASTOR MTR2, were used in the Netherlands, on a temporary basis, before the HABOG facility became available. German casks are discussed in Ref. [54].

The cask can be designed for storage purposes only, or for both storage and transport. The dual purpose cask has the advantage of being perceived as 'road-ready', to be moved to the final repository or reprocessing site without the necessity of retrieving and repackaging the stored fuel. Such casks consist of a sturdy cylindrical body with an internal cavity to accommodate the basket which holds the spent fuel elements. The cask shield wall consists of lead sandwiched between stainless steel outer and inner surfaces. The storage cavity is backfilled with helium gas, and a double lid system assures the required containment [57].

Figure 6 shows the dual purpose metal cask developed in Latin America to transport and eventually dry store spent fuel from research reactors in Argentina, Brazil, Chile, Mexico and Peru [57].

An evolution of the metal dual purpose cask is the 'flexible' dual purpose cask that is receiving considerable support in the current market of SNF management from NPPs. In this concept, the spent fuel is placed into a metal canister that becomes the spent fuel package for all future activities. This canister is first placed in a concrete overpack for long term storage. When it is time to remove the spent fuel, the canister is transferred to a reusable (metal) transport overpack for off-site shipment. Relative to the metal dual purpose cask, the flexible dual purpose casks significantly reduce the amount of metal in the system and all related costs [58].

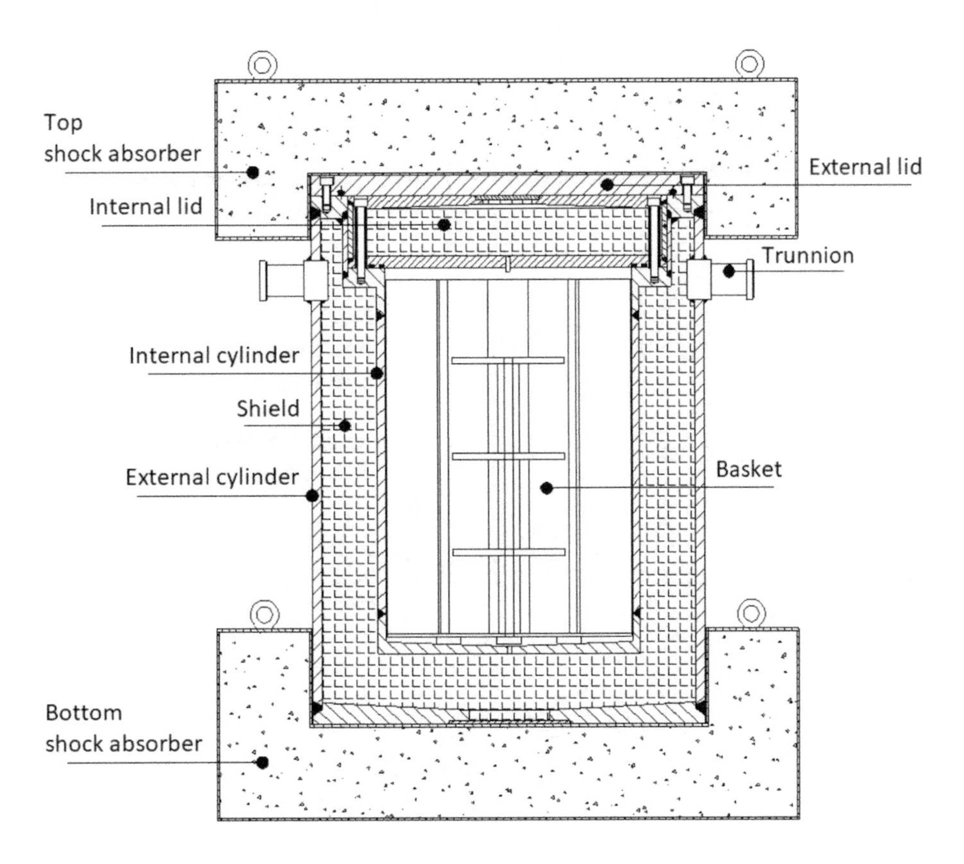

FIG. 6. Dual purpose metal casks developed in Latin America [57].

4.1.1.4. *Concrete cask*

The concrete cask (silo) system has been used in Canada to dry store spent fuel discharged from decommissioned test reactors since the 1980s. As shown in Fig. 7, the silo is an internally steel lined reinforced concrete cylinder that provides the necessary shielding and containment of the irradiated fuel. Fuel bundles are preloaded into stainless steel baskets in the pool. The baskets are then drained, dried and seal-welded before being emplaced in the silo. Each silo contains a stack of spent fuel storage baskets. The canister plug is seal-welded to the liner at the completion of all operations and IAEA safeguards seals are installed. The fuel basket and the concrete silo liner form two engineered containment boundaries for the fuel. Decay heat is transferred to the environment by conduction through the concrete walls.

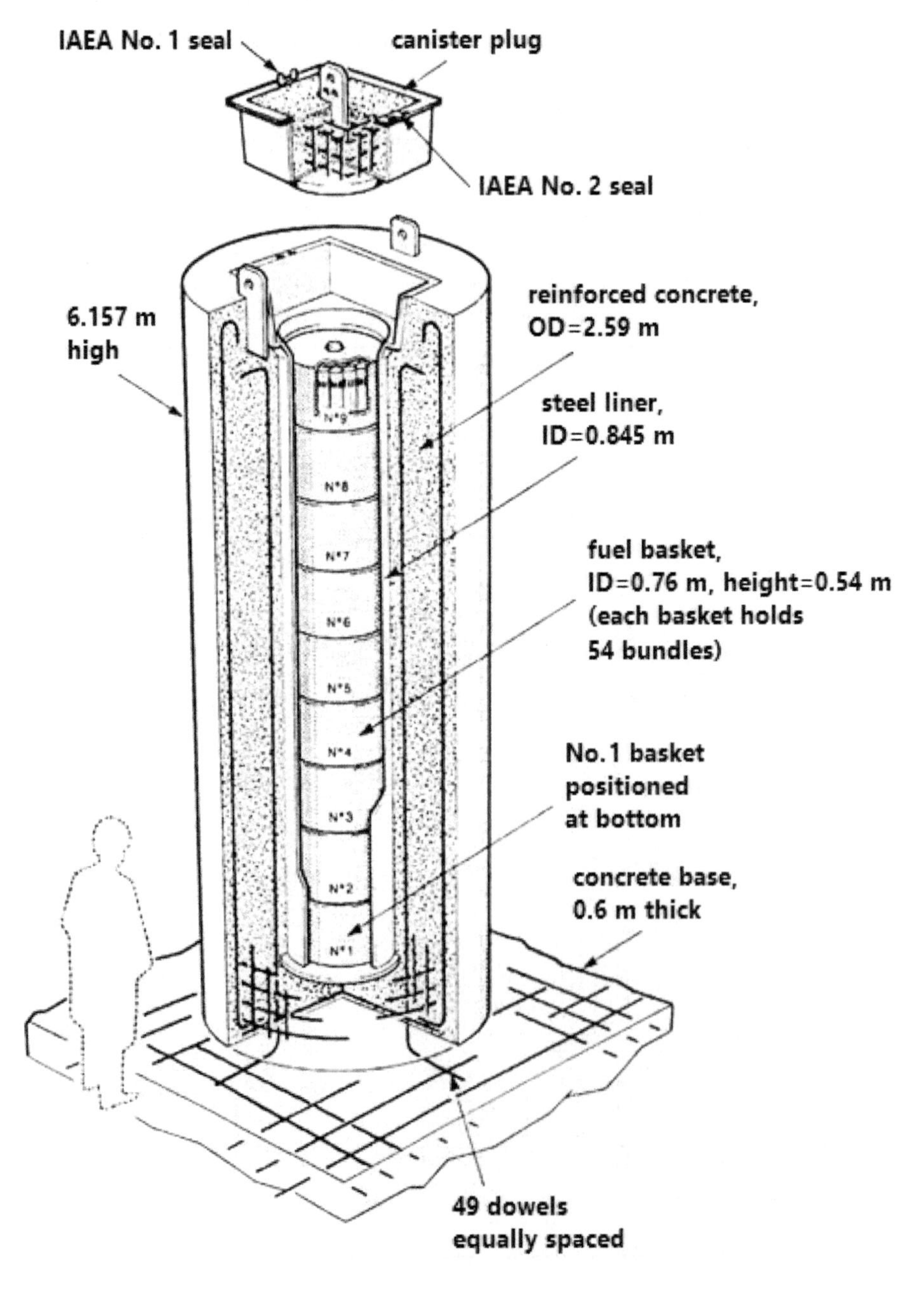

FIG. 7. Concrete silo used in Canada for SNF storage (reproduced from Ref. [146]).

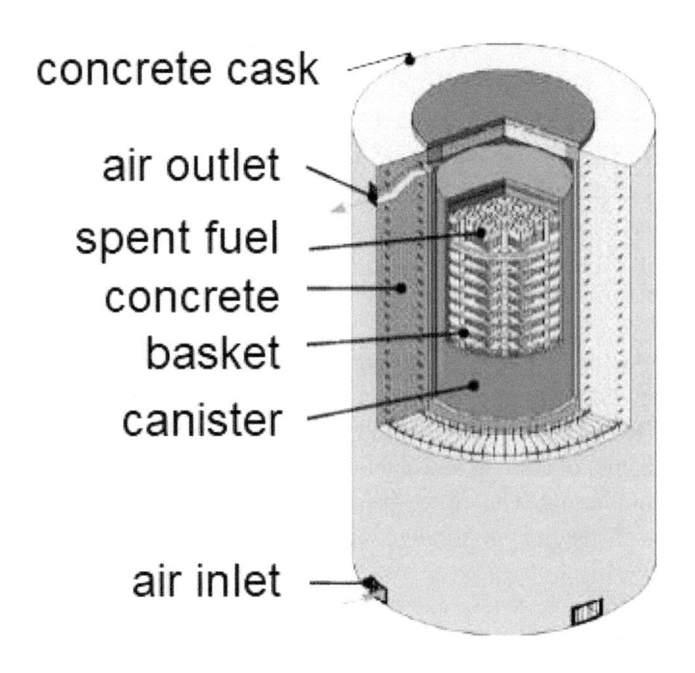

FIG. 8. Example of a ventilated concrete cask that increases the heat load for SNF storage (reproduced from Ref. [147]).

The concrete silos are a simple, inexpensive, effective dry storage design and are the least susceptible to degradation. However, given that the thermal conductivity of concrete is a factor of 10 to 40 lower than that of the metals used in metal casks, the concrete silo thermal design is acceptable only for relatively smaller total decay heat power (<4 kW). To improve the thermal performance of concrete casks, ventilated concrete casks have been designed and used for dry storage of spent fuel from commercial power reactors. Figure 8 is a sketch of a ventilated concrete cask developed in Japan. An air passageway around the storage canister in this design significantly increases the cask's heat load to about 24 kW.

4.1.2. Operational and economic considerations

Interim dry storage can be a competitive and economic alternative to interim wet storage, and may be the best option to store aluminium clad RRSNF. This technology allows the continued use of reactor facilities and can be used as either an alternative or a complementary option to interim wet storage. Sections 4.1.2.1 to 4.1.2.4 discuss the following issues:

— Operational considerations;
— Economic considerations;
— Cost for engineering and licensing;
— Capital cost for construction.

4.1.2.1. Operational considerations

Aluminium clad research reactor fuel is prone to corrosion if wet stored in a water pool of less than high purity. In dry storage, where the spent fuel is stored in a dry and often inert condition, corrosion degradation of the fuel cladding is minimized or eliminated during long term storage. This allows a much longer period of interim storage, considerably extending the time available to decide disposal options.

The distinctive characteristic of dry storage systems is the completely passive cooling, with no active components and no reliance on electricity and water. This feature makes dry storage inherently safer and more reliable than wet storage. In-pool storage, heat removal and shielding of stored fuel would be lost in the event of a loss of pool water accident. Uncovering of spent fuel assemblies could lead to fuel overheating, resulting in severe radiological consequences.

Spent fuel from research reactors contains significant quantities and concentrations of fissile materials. Subcriticality of the RRSNF has to be assured at all times during long term storage. In this regard, dry storage offers an advantage in criticality safety control over wet storage due to the absence of water, which is an excellent moderator [20].

4.1.2.2. Economic considerations

The costs associated with the design, licensing and construction of most dry storage are expected to be lower than that of a new SNF storage pool. This is due to the structural, equipment and active system requirements of a pool that also have to be enclosed in a properly qualified structure.

Due to the passive nature of cooling and shielding, dry storage requires little maintenance except for periodic inspection and monitoring. Therefore, the operating costs for dry storage are considerably lower than for wet storage, which requires continuous operation of several safety related systems (e.g. heat removal system, water purification system, makeup water system, ventilation exhaust system). The safety related systems required for wet storage include a large amount of active mechanical components, resulting in higher operational complexity and maintenance costs. In comparison with wet storage, which generates an appreciable amount of low level waste from the operation of cooling, filtration, cleaning and sampling systems, dry storage generates less low level waste, thus reducing the costs associated with waste management.

A comprehensive analysis of the life cycle costs of dry storage facilities for spent fuel from commercial power reactors is given in IAEA Nuclear Energy Series No. NF-T-3.5, Costing of Spent Nuclear Fuel Storage [58]. However, information on the cost of different SNF storage technologies is limited due to its proprietary nature. In the absence of data specific to the costs of dry storage facilities for RRSNF, Sections 4.1.2.3 and 4.1.2.4 provide a qualitative discussion of the costs associated with dry storage of RRSNF. The information given is based on a USDOE document on storage options for returned foreign RRSNF, Ref. [59].

4.1.2.3. Engineering and licensing cost

To minimize the cost and schedule for the completion of any storage facility for RRSNF, it would be preferable to choose licensed dry storage systems for commercial power reactor spent fuel (see Section 4.1.3), with modifications to accommodate the specific characteristics of the RRSNF. A good example is the CASTOR MTR2 metal casks used at the Ahaus site in Germany and the Central Organization for Radioactive Waste (COVRA) site in the Netherlands. The CASTOR MTR2 is a smaller version of the dual purpose cask concept, specifically designed for their RRSNF.

Reference [60] recommends that redesign engineering be limited to changes in the design of the basket that encapsulates the fuel. Outside the basket, all remaining components have to be identical to those already licensed for commercial nuclear fuel, thereby significantly reducing engineering analysis, license review time and costs, as well as drawing and specification changes.

4.1.2.4. Capital construction cost

The construction cost of dry storage facilities for RRSNF varies significantly between designs. In terms of the unit capital costs for different storage designs, the least expensive unit is the concrete silo due to its simple concrete–canister design, followed by the ventilated concrete cask equipped with internal air passage labyrinth. The most expensive unit is the metal cask, due to its use of a thick metal wall instead of concrete [59]. The construction cost of the drywell type designs is expected to fall somewhere between the silo and the ventilated concrete cask.

For dry storage of large quantities of spent fuel for a prolonged period, the ventilated modular dry vault is attractive. While the up-front cost of establishing a vault is substantial, the unit capital cost for dry storage of the spent fuel is expected to be competitive.

The metal casks, concrete silos, ventilated concrete casks and drywells are modular, piece-by-piece storage designs, and can be constructed or procured as needed. This flexibility can potentially provide considerable economic benefits.

All the dry storage designs mentioned above, excluding the dual purpose metal casks, are for dry storage only. This implies that, at the end of the designed storage life, the stored spent fuel will have to be retrieved and repackaged to meet the safety requirements for transport established in IAEA Safety Standards Series No. SSR-6 (Rev. 1), Regulations for the Safe Transport of Radioactive Material, 2018 Edition [61]. The dual purpose metal cask has the advantage of being perceived as 'road-ready', thus eliminating the process and costs of retrieving and repackaging the stored fuel. By minimizing fuel handling operations, the dose for workers and the volume of additional low level waste produced can be reduced. However, in selecting the dual purpose casks, the users explicitly assume that the current safety requirements for spent fuel transport [61], based on which the casks are designed, will not change significantly over the design life for spent fuel storage.

4.1.3. Licensed dry storage systems

RRSNF management is regulated by the national regulatory bodies in respective States. Therefore, a dry storage system licensed in one State may not necessarily be licensed automatically in other States. However, considerable savings in cost and schedule can be accrued by adopting, with minor modifications, a tested and proven dry storage design that is licensed in States with major nuclear programmes for commercial power reactor spent fuel, as in the case of the CASTOR MTR2 metal casks used in Germany and the Netherlands.

A comprehensive overview of dry storage technologies that have been licensed in the USA for spent fuel from commercial power reactors is provided in Ref. [62].

4.1.4. Sealed and vented technologies

One of the fundamental requirements of a dry storage design is to contain the spent fuel in order to prevent unacceptable releases of radioactive material. A multiple-barrier approach is adopted to ensure containment. A barrier is defined as a natural or engineered feature that delays or prevents material migration to or from storage components [20]. These barriers may include the fuel matrix and cladding. Further containment barriers may include a container (or storage tube or liner) as an integral part of the storage system.

Prior to placement into dry storage, the spent fuel assemblies are loaded underwater into the storage containers. The container is then dried, backfilled with inert gases and closed to ensure a dry and inert environment for the spent fuel. However, an excessive amount of water may still be present if the container and its contents are not adequately dried, or if humidity or water finds its way into the container during storage. The presence of excessive water in the container over the proposed storage period could (a) corrode the fuel assemblies and compromise the fuel integrity, and (b) present an energetic (fire or explosion) hazard owing to buildup of hydrogen gas from corrosion and radiolysis, and owing to formation of pyrophoric corrosion products (see Section 4.1.7). To address these issues, two distinct types of storage container designs are used: a sealed system and a vented system.

A sealed storage system is one in which a fully sealed container, enclosing one or more fuel assemblies, is placed within a dry storage facility. The approach is to sufficiently dry the contents in the sealed container to ensure that the amount of residual water, if fully reacted, would not be sufficient:

- — To degrade the fuel assemblies and the container material to unacceptable levels;
- — To cause an energetic event by accumulation of hydrogen gas and formation of pyrophoric substances;
- — To exceed the design pressure of the sealed container due to gas buildup.

If a sealed system is used, the seal needs to last for the entire duration of the storage period (e.g. 50 years) or allowance (e.g. cost) has to be made for resealing [60].

A vented system stores SNF assemblies in non-sealed containers or holders that are accessible to the environment of the facility. One example is the tile holes at Chalk River in Canada. Vented systems have the advantage that residual water in and around the spent fuel can evaporate or undergo radiolysis over long times and escape from the system. However, exposure of the container to ambient air also allows ingress of moisture by aspiration of humid air from the external atmosphere. For below-grade storage facilities containing cool fuels, the entered moisture would condense and accumulate in the container. In vented storage systems where cyclic water ingress is possible, galvanic coupling between the stainless steel and aluminium may be a concern with relatively cold fuel. To prevent the aspiration of humid air, the vent ports of liner tubes in the ANSTO drywells were normally kept closed mechanically, and were open only when conducting periodic inspection and sampling [53].

The technical basis for the vented design is that the corrosion of the aluminium cladding and the buildup of hydrogen gas progress at slow rates [60], allowing sufficient time for remedial actions (e.g. redrying and refilling of inert gas), if found to be necessary following periodic monitoring, inspection and testing.

4.1.5. Technical considerations for dry storage

The basic functions required for any spent fuel storage facilities under all normal, abnormal and credible accident conditions are:

- To maintain fuel in a subcritical condition;
- To remove the residual heat of the spent fuel;
- To maintain containment;
- To provide for radiation protection;
- To allow retrieval over the anticipated lifetime of the facility.

These design requirements are applicable to both wet and dry storage, and are common to both power reactor spent fuel and RRSNF. Recommendations on how to meet these design requirements (subcriticality, heat removal, shielding, containment and retrievability) are given in IAEA Safety Standards Series No. SSG-15 (Rev. 1), Storage of Spent Nuclear Fuel [20] and No. SSG-27, Criticality Safety in the Handling of Fissile Material [63]. The discussion in this section focuses on issues specific to dry storage of RRSNF.

Compared with Zircaloy clad power reactor fuels, aluminium clad research reactor fuels are significantly less robust, less corrosion resistant, and are subject to unique hazards associated with their chemical and physical characteristics. Corrosion of the cladding and fuel occurs from exposure to air and sources of humidity. The presence of excess humidity or water in the fuel container would result in excessive corrosion of the fuel assemblies during storage, leading to:

- Poor retrievability due to loss of fuel integrity and excessive distortion of fuel assemblies;
- Impaired containment as a result of cladding breach;
- Risk of fire or explosion due to generation of hydrogen gas from fuel corrosion and water radiolysis, and generation of pyrophoric materials;
- Increased risk of loss of subcriticality due to the presence of water and loss of the original configuration of fuel assembly.

For this reason, the vulnerability to degradation and associated retrieval difficulties are considered to be the two major issues specific to dry storage of RRSNF.

4.1.5.1. *Safety considerations*

A gross rupture of the spent fuel cladding due to excessive degradation or an energetic event (fire or explosion, or over-pressurization of a sealed container) in the storage system has to be prevented to maintain the primary confinement barrier of the fuel clad system and to ensure the post-storage retrievability. The degradation mechanisms of the fuel and the potential hazards that can lead to severe damage to the storage system are discussed in Sections 4.1.5.2 and 4.1.5.3.

4.1.5.2. *Degradation of the fuel aluminium cladding*

Reference [64] provides an evaluation of the potential degradation mechanisms of aluminium clad spent fuel during dry storage. The range of environmental conditions considered are: temperatures less than 600°C; cover gases of air, non-reactive, or inert gas; and relative humidity levels up to 100%. The potential degradation mechanisms identified and evaluated include:

— Corrosion of the cladding and fuel from exposure to air and sources of humidity;
— Dehydration of the hydrated oxide films with subsequent corrosion of the cladding;
— Hydrogen blistering of the aluminium cladding;
— Radiation embrittlement of the cladding;
— Interdiffusion of the fuel and fission products with the cladding;
— Creep of the cladding due to self-weight loading and loading from swollen fuel.

In Ref. [65] it was concluded that these mechanisms, except for corrosion, are only significant at temperatures well above 200°C. Given this, for dry storage facilities of aluminium clad spent fuel, the maximum fuel temperature should be 200°C to prevent occurrence of the above degradation mechanisms other than corrosion. Therefore, in the presence of water vapour or air, corrosion of the aluminium cladding and the fuel matrix materials is the most limiting mechanism that could cause significant degradation of the fuel during dry storage.

The rate of corrosion increases with the presence of elevated levels of impurities (e.g. chloride and sulphate ions) in the water vapour. Gamma radiolysis of air produces various gaseous species, including nitrogen oxide gases. When moisture is present, nitric acid can be produced to accelerate the corrosion.

4.1.5.3. *Potential hazards of severe damage to the fuel storage container*

Internal events, such as fire, explosion, or over-pressurization of a sealed container can result in severe damage to the fuel storage system, including fuel storage containers. For a dry storage facility containing aluminium clad RRSNF, the potential hazards that can lead to such damages are: (a) fire and explosion due to hydrogen buildup or sealed container over-pressurization due to gas generation; and (b) fire and explosion due to formation of uranium hydride.

4.1.6. Hydrogen buildup

At temperatures above approximately 80°C, hydrogen buildup occurs in a closed system containing aluminium and water due to the following chemical reaction [65]:

$$2\,Al + 4\,H_2O \rightarrow Al_2O_3 \bullet H_2O + 3\,H_2$$

The generation of hydrogen in the reaction to produce boehmite (AlOOH or $Al_2O_3 \bullet H_2O$) bounds that for the reaction to produce gibbsite ($Al(OH)_3$ or $Al_2O_3 \bullet 3H_2O$) which occurs at temperatures below approximately 80°C [64].

One impact of hydrogen buildup is the potential for an explosion hazard. The lower concentration limits are:

— 4% by volume in air at room temperature for flammability of hydrogen;
— 9% by volume in air for a sustained burn of hydrogen;
— 18% by volume for hydrogen and air mixture being explosive.

The lower limit of 4% by volume is applied as the limit to the amount of hydrogen that can be generated in the fuel storage containers. The maximum allowable amount of free water (W, in grams) is expressed as a function of the free volume (V, in cm^3) of the container [64]:

$$W = 3.873 \times 10^{-5} \times V$$

In other words, no more than 38 grams of free water is allowed in one cubic metre of free volume of the fuel storage container to ensure the hydrogen concentration due to aqueous corrosion of aluminium is below the flammability limit [66].[4]

If an inert gas is used as the cover gas for the fuel storage container containing dried aluminium clad fuel, the potential to produce a flammable mixture would be essentially eliminated.

In addition to being a fire or explosion hazard, accumulation of excessive hydrogen gas may also over-pressurize the sealed storage container and cause hydrogen embrittlement of container materials at locations of high stress and surface discontinuities over long storage times [42].

4.1.7. Generation of pyrophoric uranium hydride

When a breach of the aluminium cladding occurs, the fuel core material is exposed to the corrosive storage environment. For RRSNF with uranium metal cores, formation of uranium hydride (UH_3) is a potential safety issue.

Uranium metal is chemically very reactive, with both water (liquid or vapour) and every major atmospheric component [6]. The reaction with water is considered the primary cause of corrosion of the exposed uranium metal under conditions relevant to dry storage. The products of a highly exothermic uranium–water reaction are uranium dioxide (UO_2) and hydrogen gas. Under oxygen-depleted and hydrogen-rich conditions, such as those in a sealed storage container, the molecular hydrogen generated is dissociated by uranium and forms UH_3. The product, a loose fine black particulate, is pyrophoric at room temperature when exposed to air.

This presents a potential explosion hazard for fuel retrieval at the end of interim dry storage. The loose UH_3 powder would be disturbed and may become suspended in the air when the fuel storage container is handled during retrieval. When the fuel container closure is opened, a violent reaction of the oxygen in the air with the suspended UH_3 may cause an explosion [67].

Due to the high reactivity of the uranium metal, fuel with exposed uranium metal would need to be thoroughly dried and stored in dry helium cover gas.

4.1.8. Retrievability considerations

In most cases when dry storage is considered, interim storage containers are not foreseen to be used for disposal. Therefore, at the end of storage life, the RRSNF would need to be retrieved safely from interim dry storage for re-conditioning and repackaging prior to its placement in the disposal facility.

---

[4] When expressed as the ratio of residual water content to net free volume, this dryness limit is equivalent to a ratio of 3.8×10^{-5}. This dryness limit appears very tight and may be difficult to meet. For example, the USDOE standardized canister allows a much higher residual water to free volume ratio of 8.7×10^{-3} after drying (see Ref. [67]).

Retrievability of the stored spent fuel is related directly to limiting degradation of the storage system. Excessive degradation would distort the configuration and reduce the mechanical strength of the stored fuel assemblies to a degree that may compromise its retrievability.[5]

4.1.9. Dry storage acceptance criteria and operational envelope

An extensive research and development programme has been conducted at the USDOE SRS since the early 1990s to develop regulatory acceptance criteria with technical basis for dry storage of aluminium clad RRSNF over 50 years. The acceptance criteria are the combined set of environmental limits for drying and for storage to avoid excessive degradation, to maintain a primary confinement barrier of the fuel clad system and to ensure the post-storage retrievability during the nominal 50 year storage period envisioned. The drying criteria specify the allowed maximum fuel temperature during the drying process and the amount of residual water in the fuel container after drying. The storage criteria define the operational envelope during dry storage, such as allowed maximum fuel temperature, type of cover gas, etc. These criteria vary with the types and conditions of the RRSNF, as well as the types of selected dry storage system (i.e. sealed or vented storage).

The acceptance criteria proposed by USDOE Savannah River Laboratories for interim storage of the major types of RRSNF at the SRS are summarized in Table 3 [43].[6]

TABLE 3. ACCEPTANCE CRITERIA FOR DRY STORAGE OF ALUMINIUM CLAD CATEGORY 1, 1A AND 2 FUEL MATRIX MATERIALS

Fuel category	Fuel condition	Storage system	Drying criteria		Storage criteria	
			Max. drying temperature	Max. free water in sealed system	Storage environment	Max. cladding temperature
Categories 1 and 1A	Intact	Sealed system	Up to 250°C for up to 2 hours	1 mL free water per 0.1 m^2 of cladding surface	Non-reactive gas, air or vacuum acceptable	200°C
		Vented system	Up to 250°C for up to 2 hours	n.a.	Flowing ambient air required	200°C
	Breached	Sealed system	Up to 200°C for up to 2 hours	1 mL free water per 0.1 m^2 of cladding surface	Non-reactive gas, air or vacuum acceptable	200°C
		Vented system	Up to 200°C for up to 2 hours	n.a.	Flowing ambient air required	200°C

[5] Excessive degradation of the storage container would also affect retrievability. However, as the container materials (typically stainless steel) are less vulnerable to degradation during dry storage owing to their higher corrosion-resistance and mechanical strength, degradation of the aluminium clad RRSNF is more limiting.

[6] Flowing ambient air is proposed in Ref. [46] for vented system to sweep N_xO_y gases generated from air radiolysis. In some of the vented designs for storage of RRSNF with lower decay heat power (e.g. ANSTO in Australia), the storage container is backfilled with inert gas but vented and refilled periodically via a normally closed vent port.

TABLE 3. ACCEPTANCE CRITERIA FOR DRY STORAGE OF ALUMINIUM CLAD CATEGORY 1, 1A AND 2 FUEL MATRIX MATERIALS (cont.)

Fuel category	Fuel condition	Storage system	Drying criteria		Storage criteria	
			Max. drying temperature	Max. free water in sealed system	Storage environment	Max. cladding temperature
Category 2 (U metal)	Intact	Sealed system	Up to 250°C for up to 2 hours	1 mL free water per 0.1 m^2 of cladding surface	Non-reactive gas, air or vacuum acceptable	200°C
		Vented system	Up to 250°C for up to 2 hours	n.a.	Flowing ambient air required	200°C
	Breached	Sealed system	Up to 200°C for up to 2 hours. Dry under non-reactive gas only	1 mL free water per 0.1 m^2 of cladding surface	Non-reactive gas, air or vacuum acceptable	200°C
		Vented system	Not recommended	n.a.	Not recommended	n.a.

Notes: n.a.: not applicable, max.: maximum.

Category 1 is fuel with Al–uranium alloy fuel matrix and aluminium cladding. Category 1A is fuel with Al–uranium silicide or Al–uranium oxide fuel matrix and aluminium cladding.

Category 2 is aluminium clad fuel with core materials of uranium metal or other uranium alloy fuel that is not a stable uranium compound dispersed in a matrix of aluminium.

The free water limit for a sealed system specified is based on limiting the predicted corrosion of the cladding and exposed fuel to well within 7.62 µm (0.0003 inch) in depth over the 50 year storage period.

The maximum cladding temperature of 200°C during storage would ensure that most of the potential degradation mechanisms, other than corrosion, would be insignificant.

Breached uranium metal fuel (Category 2) may need stabilization and additional canning before placement in a sealed container with inert cover gas.

4.2. DRY STORAGE PACKAGE DESIGN CONSIDERATIONS

4.2.1. Material selection

The materials selected for the dry storage system need to have adequate corrosion performance during normal operation, abnormal operation and accident conditions in the environmental conditions of the facility for the duration of the design life. Thus, materials of structures and components in direct contact with spent fuel assemblies need to be compatible with the materials of the fuel assemblies [20].

4.2.1.1. Canister, basket and SNF compatibility

Stainless steel is typically selected as the material for the canister and fuel basket due to its excellent corrosion resistance and mechanical properties. Under typical dry storage conditions, stainless steel coupled to aluminium would not lead to significant corrosion [43]. The stainless steel is compatible with a canister interior composed of stainless steel and aluminium components, including various neutron source assemblies [46].

The type of the stainless steel needs to be compatible with the surrounding atmosphere, for instance 304 L for a neutral atmosphere or 316 L for an aggressive atmosphere at coastal marine locations or areas with industrial air pollution.

4.2.1.2. Canister design life

The design life of the canisters for storage of SNF from research reactors and commercial power reactors is normally 40–50 years. Due to the unknowns in the schedule for the end point of storage (i.e. reprocessing or disposal), storage designs have to be made as flexible as possible, with provisions for facility lifetime extension. The canister design has to ensure that the safety functions (subcriticality, residual heat removal, shielding, containment and eventual retrievability) are maintained under the environmental conditions expected during its design life [20].

The materials used in a dry storage system are subject to adverse alteration processes (degradations) under long term storage conditions. These alteration processes have to be evaluated. An acceptable methodology for lifetime prediction will (a) identify alteration mechanisms; (b) quantify the rate of alteration for each mechanism; (c) evaluate the effects (on material properties and condition) of each alteration; (d) determine if the alterations compromise any safety functions of the system; and (e) determine the consequences of compromising the performance of the component, the subsystem or the system (safety, operational, economic) [68].

4.2.2. Canister and package design

4.2.2.1. Sealed and vented designs

Maintaining a truly dry and non-reactive storage environment is the key issue for safe and retrievable dry storage of RRSNF. The sealed and vented canister designs differ in their approaches to address this issue.

In a sealed canister design, a loaded canister is dried, seal-welded, tested for leaks, and backfilled with inert gas. A dryness criterion needs to be established to ensure that the amount of residual water after drying would not cause unacceptable degradation of the contents or unacceptable risks of an energetic event or over-pressurization of the canister. Therefore, this design requires the justification of the specified dryness criterion, the demonstration of meeting the dryness criterion, and assurance of the functionality of the sealing system to maintain this dryness level over its design life. In other words, a sealed canister design places a great emphasis on achieving and maintaining a non-reactive and sufficiently dry environment for the spent fuel stored. The canisters need to be designed as a pressure-containing vessel to accommodate potential pressure buildup due to the presence of residual water or humidity after drying. Most of the existing dry storage designs for RRSNF, including the modular vault, metal cask and concrete cask, are a sealed design.

The dryness criteria are often very tight; achieving this dryness level can be a significant technical challenge as well as a lengthy and costly process. To demonstrate to the regulator that the amount of the residual water or humidity after drying is actually below the dryness limit may also be a difficult task.

An alternative design is the non-sealed vented system in which the stored fuel is accessible to the atmosphere. After the fuel container is loaded with spent fuel, the container is dried to remove as much free water as practical. The container is then mechanically sealed and backfilled with inert gas. Unlike the sealed design, the vented container is equipped with a vent port on top. The vent port is normally closed by a lid or an isolation valve, and is vented periodically for monitoring, sampling, re-drying and refilling. Figure 9 shows the vent (sampling) port and the gas sampling system used at the drywell storage facility of ANSTO, Australia.

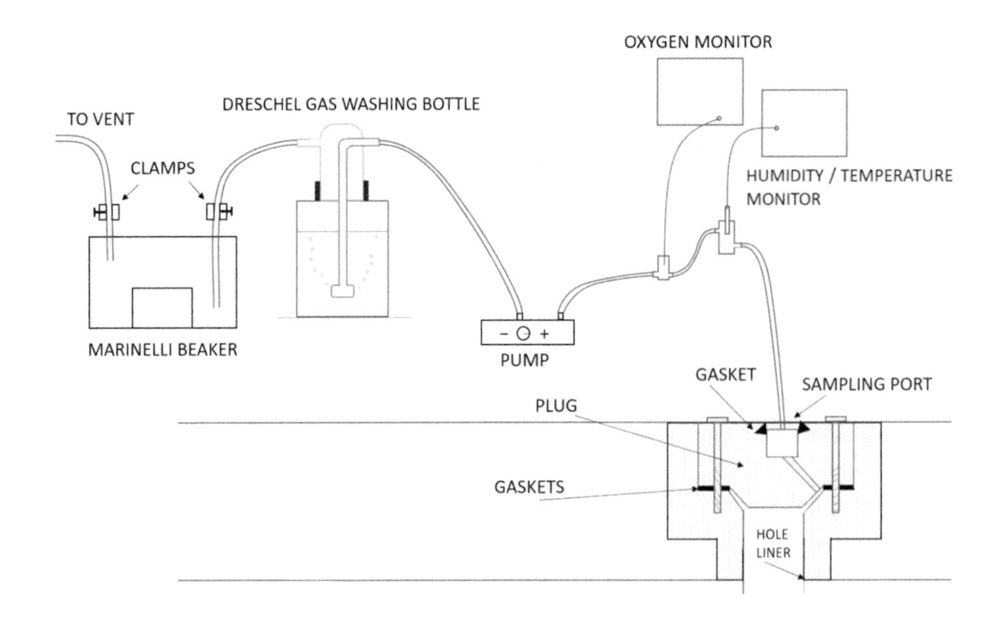

FIG. 9. Vent (sampling) port of the drywells in ANSTO.

The vented canister design does not rely on the drying process to achieve a high level of dryness. Instead, it accepts that some forms of the residual water cannot be removed completely. These forms of residual water, if released, would corrode the fuel assemblies and generate hydrogen gas. However, the fuel corrosion and hydrogen buildup in the canister is a slow process and can be accommodated by periodic gas monitoring. Higher pressure and hydrogen concentration in the gas indicates the presence of excessive residual water. In situ gas purging, redrying and refilling capability need to be provided for this particular storage location with a portable dryer on a skid. The vented canister needs not be designed as a pressure vessel.

The vented system would require more monitoring and inspection operations and, hence, higher radiological exposure for the facility staff than a sealed system would. It is important that the design of the vented system facilitate the monitoring, purging, redrying and inert gas refilling process to minimize exposure of the staff.

4.2.2.2. Subcriticality, shielding and thermal considerations

Subcriticality, shielding and thermal parameters are the most important design considerations for dry storage systems. As an extensive discussion of these design requirements is beyond the scope of this publication, only a brief discussion is provided below. Recommendations on how to meet these design requirements for spent fuel storage facilities can be found in SSG-15 (Rev. 1) [20].

The RRSNF contains significant quantities and concentrations of fissile materials, and subcriticality of the RRSNF has to be assured at all times during long term storage. Subcriticality can be ensured by different means, including [59, 63]:

— Diluted concentration of fissile materials;
— Adequate separation distance between fuel assemblies;
— Presence of neutron absorbing materials (such as boron) mixed with the fissile material;
— Exclusion of substances such as water that are good moderators and reflectors for neutrons;
— Restricting the mass of fissile material below the critical mass.

An SNF storage system has to provide adequate shielding for both the gamma and neutron radiation that emanates from irradiated nuclear fuel. The shielding has to be designed in such a way as to reduce the combined gamma and neutron dose rate to values that are below the limits set by the respective regulatory bodies for the public and the on-site workers.

Dry storage technology relies on a number of solid shielding materials, sometimes in combination, to reduce gamma and neutron dose rates. The most common materials used for shielding are concrete, cast iron, carbon, stainless steel, lead, borated resin and polyethylene. The material selection depends on design limitations regarding shield thickness, cost, strength, weight and safety. Lead and steel can provide relatively more effective gamma shielding with thinner material. However, using steel or lead imposes a design penalty of increased cost. Concrete is much less expensive and may reduce overall shielding costs.

The dry storage technologies rely on a passive heat removal system to ensure that the temperatures of the spent fuel assemblies as well as the components of the storage system are maintained below their respective limits appropriate for long term storage. For the aluminium clad RRSNF, a limit of 200°C is recommended for the spent fuel cladding during dry storage (see Section 4.1.8).

In a metal cask, the fuel decay heat is transferred to the canister's walls, then to the metal cask wall. At the outside of the metal cask, the heat is removed by conduction and natural convection to the environment. Some designs incorporate metal fins on the exterior of the cask to enhance heat transmission to the air. The metals used in SNF storage cask designs are ductile cast iron, carbon steel, lead, and stainless steel. In terms of their heat conduction properties, cast iron, lead and carbon steel are superior to stainless steel because they have a thermal conductivity that is about three times that of stainless steel. The metal cask heat transfer system is not susceptible to thermal limits, since these metals have a higher temperature limit than fuel cladding [59].

Concrete is used as the shielding and structural material for the modular concrete vaults and the concrete casks. As the thermal conductivity of concrete is a factor of 10 to 40 lower than that of the metals used in a metal cask, a labyrinth airflow passage design is often required to accommodate higher decay heat load.[7] The labyrinth airflow passage allows natural convection-driven air to enter the cavity enclosing the canister inside the concrete, and then exits to the environment through the air outlets at a higher elevation (see Fig. 8). The need for these airflow passages and their associated inlets and outlets introduces the possibility of an accident in which the inlet and outlets could be blocked by debris, snow, or even nests or hives. Therefore, concrete casks require surveillance of their air inlet and outlet flow passages.

In addition to shielding, reinforced concrete may also perform a structural function. In this case, the concrete temperature needs to be maintained below certain limits. Recommended concrete temperature limits are provided in Ref. [68]. For dry storage systems that use thick concrete walls, temperature distribution within the concrete structure is also important, in addition to the maximum concrete temperature. High temperature gradients may cause excessive thermal stress and degradation of the concrete structure.

The heat removal from a dry storage system can also be improved by selecting a cover gas with higher heat conductivity in the fuel storage container or canister, as discussed in Section 4.2.2.3.

4.2.2.3. Backfill gas selection

After a loaded fuel canister is dried, the cavity of the canister is backfilled with non-reactive or inert gases for the following purposes:

— To stabilize the stored fuel;
— To eliminate or slow the formation of an explosive gas mixture and pyrophoric substances;
— To enhance heat transfer by introducing cover gases with higher thermal conductivity.

The most commonly used cover gases are helium and argon. As helium gas has a thermal conductivity about nine times higher than that of argon gas, helium is a preferred cover gas for dry storage of spent fuel with a higher decay heat power. Nitrogen and dry air are also used in some dry storage facilities. However,

[7] One exception is the silo design which removes decay heat only by concrete conduction.

in the event of improper drying or leaks in the canister, moist air or free water may inadvertently be present in the sealed canister. Radiolysis of residual moisture in the presence of nitrogen impurities may form weak solutions of nitric and nitrous acids which attack aluminium cladding aggressively.

The cover gas may contain oxidizing impurities, including oxygen, carbon dioxide, carbon monoxide and residual moisture. The sources of the oxidizing impurities, their removal and their effects on dry storage of light water reactor spent fuel are discussed in detail in Ref. [69] which recommends a limiting maximum quantity of oxidizing gases of 0.25 vol. %. Similar assessment of the impacts of cover gas impurities on the aluminium cladding of RRSNF has yet to be done. It is expected that the impurities may have more pronounced impacts on the less corrosion resistant aluminium cladding of RRSNF, driving the impurity limit to a more restrictive level.

4.2.2.4. Closure technology

For a vented storage design, such as the drywell stores shown in Fig. 9, the fuel storage tubes are closed by shield plugs. The shield plug is bolted and tightened by means of a gasket to the flanged top of the storage tube.

A sealed canister or package design can be closed with use of a bolted closure or a weld sealed closure. All operations involved in the canister closure have to be performed remotely to minimize the dose to workers. The sealing system is designed to withstand all postulated accidents and maintain integrity over the lifetime of the cask, because it constitutes part of the radioisotope confinement boundary.

Dry storage casks and canisters with bolted closures involve configurations that have metal or elastomer gaskets. Figure 10 shows the closure system of a CASTOR cask used at the Ahaus site in Germany [54]. The storage cavity is closed with double lids made of stainless steel. A primary lid closes the cask cavity and a secondary lid is bolted to the cask body above it. Bolted lids are equipped with long-lasting, high efficiency, metallic seals. The space between the lids is filled with helium at 600 kPa (6 bar). This pressure is continuously monitored during the storage period as a reliable indication of seal integrity.

In case the second lid leaks, it can be repaired or replaced without affecting the inner primary containment. For changing a seal on the primarily lid, the cask would have to be shipped to a special facility in which the cask is opened and the corresponding gasket is replaced. Alternatively, the cask concept can include repair by placing a third lid above the secondary one, thereby re-establishing the double barrier.

Welding is the more popular method used to seal canisters (e.g. the MCO used at the Hanford site in the USA). Strict quality control during welding and a periodic inspection of weld integrity is needed. Leak tightness of the welds may be verified by hydrostatic pressure tests and specific NDEs (e.g. ultrasonic testing or radiographic testing).

Compared with the welded closure, the bolted closure design offers the advantage of relative ease for cask opening and access to the stored fuel assemblies.

4.2.2.5. Certification and licensing requirements

The certification and licensing requirements for a dry storage system of RRSNF may vary from country to country. Reference [70] provides details of the licensing requirements for an independent storage of SNF in the USA.

As discussed in this Section, different designs of spent fuel dry storage facilities are operational. Each application for certification or licensing of a spent fuel dry storage facility includes a safety case[8], which provides (i) a full description of the SSCs of the spent fuel storage facility; (ii) the applicable

[8] The safety case is a collection of arguments and evidence in support of the safety of a facility or activity. This collection of arguments and evidence may be known by different names (e.g. safety analysis report or safety dossier) in different States and might be presented in a single document or a series of documents.

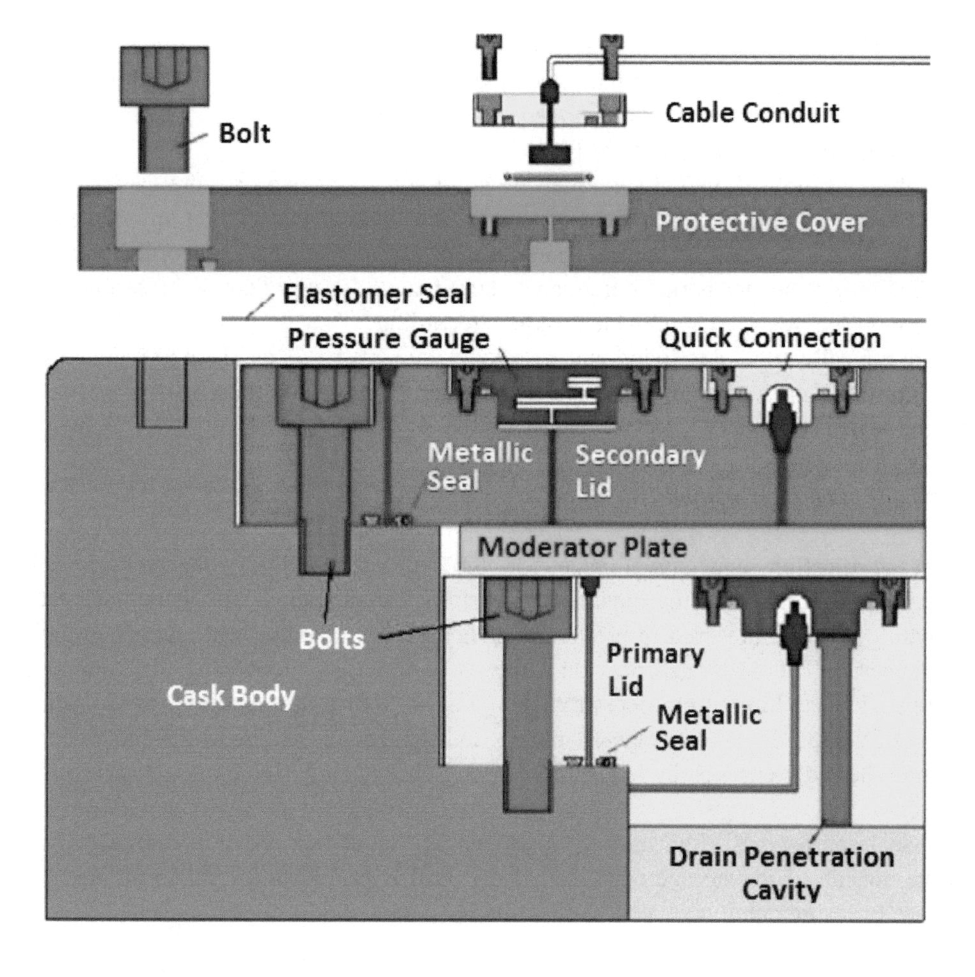

FIG. 10. Closure system of a metal cask (reproduced from Ref. [54]).

performance criteria; and (iii) a general description of the operation of the proposed facility. The purpose of a safety case is to demonstrate that a spent fuel storage facility can be implemented in compliance with the safety criteria defined by laws or regulations valid in the country. More details of preparing the safety case are provided in SSG-15 (Rev. 1) [20].

4.2.2.6. Safeguards considerations

RRSNF contains certain amounts of plutonium (buildup during reactor operation) and, therefore, has to be appropriately secured and safeguarded.

'Safeguards' refers to the IAEA safeguards system, the objective of which is the timely detection of any diversion of significant quantities of nuclear material from peaceful nuclear activities to the manufacture of nuclear weapons, other nuclear explosive devices, or for purposes unknown, and deterrence of such diversion by the likelihood of early detection. The IAEA safeguards system is based on the use of materials accounting as a safeguards measure of fundamental importance, with containment and surveillance as major complementary measures [71].

4.2.2.7. Content verification tools and methods

In a dry storage system, once the spent fuel is loaded into the container and the container is weld sealed or bolt sealed, the fuel is inaccessible until the cask is opened again for retrieval at the end of the storage period. This makes safeguarding dry storage relatively simpler than for wet storage. However, since the fuel is also inaccessible for regular materials accountancy inspection, it is necessary to rely on containment and surveillance to ensure that, once the fuel has been verified on loading into the storage system, it remains there and has not been removed. The two main containment and surveillance categories are optical surveillance and sealing systems [71].

A typical application of optical surveillance would consist of two or more cameras positioned to completely cover the storage area. The field of view of the cameras is such that any movement of items that could relate to the removal of nuclear material is easily identified and recorded as an image. The image recording may be set at a periodic frequency (significantly shorter than the fastest possible removal time) or by motion (i.e. scene change) which triggers recording.

Seals are typically applied to individual storage containers. A seal can help to indicate that material was neither introduced into nor removed from a container. At the same time, sealing provides a unique identity for the sealed container.

4.2.2.8. Tamper indication methods

Content verification and surveillance for dry storage facilities depend primarily on the tamper-resistant seals applied to the outside of the storage container. Seals, sometimes referred to as tamper indicating devices, are used to provide evidence of any unauthorized attempt to gain access to the secured material. The seals also provide a means of uniquely identifying the secured containers. Seals do not provide any kind of physical protection, nor were they designed to provide such protection. Initial loading of the fuel into the storage containers and subsequent sealing of the containers are closely monitored and repeatedly checked by IAEA inspectors. Subsequent periodic inspections confirm that the seals are intact.

A dual-seal approach is often used with dry storage containers. In this approach, two seals are applied to the outside of the storage container. If one seal is accidentally broken or fails, one more is always in place to ensure continuous verification that the container has not been opened.

4.3. DRY STORAGE SYSTEM INFRASTRUCTURE

Design, operation, and safety considerations for spent fuel dry storage facilities are discussed in SSG-15 (Rev. 1) [20]. In the following Sections 4.3.1 to 4.3.3, only a brief discussion of the key aspects is provided.

4.3.1. Facility surveillance and maintenance

Surveillance and periodic testing are used to verify that storage SSCs continue to function. Maintenance activities, both administrative and technical, are used to keep SSCs in good operating condition, including both preventive and corrective (or repair) aspects. Appropriate inspection and surveillance programmes need to be in place throughout the entire storage period [20].

Maintenance, testing and repairs have to be possible under all circumstances with respect to acceptable doses (i.e. as low as reasonably achievable). Maintenance has to be performed by qualified people in accordance with a quality assurance programme. Periodic reports on the maintenance of the installation, when applicable, have to be supplied to the competent authorities for assessment. Replacement or changing systems or components related to safety needs to be approved by the competent authorities. After installation, maintenance systems need to be tested before active service.

4.3.2. Lifetime management and in-service inspection

Interim storage facilities are mostly designed for long term (e.g. up to 100 years) storage of RRSNF. A lifetime or ageing management programme for long term operation of spent fuel storage facilities is essential. An ageing management methodology for a spent fuel dry storage facility may include:

- Screening and identifying the SSCs important to the safety of the facility;
- Identifying the principal ageing mechanisms for each of the SSCs important to safety;
- Establishing acceptance limits for ageing effects to ensure that the functionality of the SSCs important to safety is not impaired during the design life of the facility;
- Specifying an operational envelope to control and minimize the rate and effects of the identified ageing mechanisms during storage;
- Developing an effective in-service inspection and maintenance programme to monitor the conditions of the SSCs important to safety, and to provide early detection and mitigation of any abnormalities or degradation.

Detailed information on ageing mechanisms for typical materials used in dry storage systems is provided for example in Refs [6, 69, 72, 73].

In-service inspection may include examination, observation, measurement or testing undertaken to assess SSCs and materials, as well as operational activities, technical processes, organizational processes, procedures and personnel competence.

Lifetime management is the integration of ageing management with economic planning, which is undertaken for these reasons:

- To optimize the operation, maintenance and service life of the SSCs;
- To maintain an acceptable level of performance and safety;
- To maximize the return on investment over the service life of the facility.

4.3.3. Lifetime extension beyond design life

Dry storage facilities are typically designed with considerable safety margins. For example, the initial values of the decay heat and radioactive inventory of the spent fuel to be placed into dry storage are used as the basis to determine the heat removal and shielding requirements for the storage facility, and to assess the degradation of the fuel and other key components of the storage system over the entire dry storage period. The design life of the facility is then determined, often in combination with conservative assumptions of the 'worst case' site environmental conditions.

Both the decay heat and the radioactive inventory of the spent fuel decrease significantly during storage, resulting in lower stress on the storage system than predicted in the design phase. With a good monitoring, inspection, maintenance and ageing management programme, an existing dry storage facility may remain in a sound condition even at the end of its operating lifetime. In this case, lifetime extension may be considered.

Extending the storage period may require additional investigations related to both the spent fuel and the storage systems, including re-evaluation of the initial design, operation, safety assessment and any other aspect of the spent fuel storage facility relating to safety. Additionally, it may be necessary to assess the integrity of the spent fuel. Potential problems with the spent fuel integrity have to be considered in advance of the need for physical actions, such as placement into new containers. In some cases, it may be justified to move the spent fuel into a more robust storage facility rather than replacing the storage container [20].

4.4. OPERATIONAL CONSIDERATIONS FOR DRY STORAGE

4.4.1. Container surveillance

As maintaining a dry and non-reactive environment is the key requirement for the safe dry storage of RRSNF, a periodic surveillance programme of the storage containers will enable the operator to monitor the storage conditions and to take timely and appropriate corrective actions if required. The surveillance methods depend on the storage container designs.

For vented storage container designs, the relatively easy access to the internal container allows multiple surveillance methods and corrective actions to be performed. For example, in the drywell storage facility at ANSTO in Australia, the nitrogen cover gas in each of the storage wells was vented and sampled on an annual basis. The gas samples were analysed for humidity, oxygen and ^{85}Kr. A higher humidity indicated water or moisture leaking into the well. A higher level of both humidity and oxygen contents indicated degradation of the top seal gasket, while the presence of ^{85}Kr indicated a breach of the fuel cladding. All the wells were then re-dried and backfilled with a new charge of nitrogen [56]. In addition, the stored fuel assemblies from a few randomly selected wells were retrieved periodically and subjected to detailed condition assessment.

For sealed container designs with bolted closures, a seal monitoring system is needed to adequately demonstrate that seals can function and maintain a non-reactive atmosphere in the container for the storage period. Dry storage casks and canisters with bolted closures involve configurations with double long-lasting, high efficiency metallic seals. The general design approach is to pressurize the region between the redundant seals, with a non-reactive gas, to a pressure greater than that of the container cavity and the atmosphere. This pressure is continuously monitored during the storage period as a reliable indication of seal integrity. A decrease in pressure between these seals indicates that the non-reactive gas is leaking either into the cask cavity or out to the atmosphere. Packages with bolted closures can be opened in an appropriate facility (e.g. hot cell) where the stored RRSNF is taken out of the package for inspection or examination.

For sealed container designs closed entirely by weld, a continuous container monitoring system may not be required. However, the lack of a closure monitoring system is typically coupled with a periodic surveillance programme that would enable the operator to take timely and appropriate corrective actions to maintain safe storage conditions after closure degradation [46]. The frequency of the surveillance may depend on the way the packages are stored and the type of the packages. A surveillance programme that includes periodic monitoring of a statistically significant sample of the cask or canister condition is recommended. Inspection of the RRSNF itself requires a hot cell facility with a fuel transfer system adequate for the specific storage system and is not generally recommended.

4.4.2. Documentation and record retention

As the operation and maintenance of a spent fuel dry storage facility will often last for several generations, key data on the facility have to be available to make an informed decision about technical options for spent fuel management at the end of the storage period. Data for spent fuel are produced and collected in all stages of spent fuel management. Generic issues in spent fuel data management are the following [74]:

- — To identify which data are needed and the associated level of detail;
- — To identify and document procedures for extracting the data needed for spent fuel management options that may be implemented in the future;
- — To transfer such data to recording media with high longevity;
- — To establish the interfaces between the different data sets.

Some examples of the necessary information about a dry storage facility include documentation and records of [20, 74]:

— Design (e.g. design drawings and specifications);
— Construction and manufacturing;
— Modifications to the facility and equipment during design, construction, commissioning and operation;
— Maintenance, inspection and testing;
— Safety case or safety analysis report;
— Complete inventory and accounting of fuel, including characteristics and storage location;
— Control and recording of internal movements between different storage areas, as permitted by the operational limits and conditions;
— Operating procedures, including limits and conditions;
— Quality assurance and audits;
— Safeguards and physical protection.

The documentation and records should contain as much data as reasonably possible. For example, the characteristics of the stored fuel should include all production data of the fresh fuel, time in the reactor, burnup, data of the unused and produced fissile material and fission products, cooling time in the pool, etc. The data have to provide sufficient details for the dry storage period, such as packaging method, drying, backfill gas, pressure inside or outside the package or canister, canister materials and storage conditions. In principle, every piece of data that can be obtained may be useful, given that regulatory and technical requirements for transport and disposal may change.

The documentation and records need to be stored in a safe and redundant way, such as hard copies and digital copies stored at different places. The data have to be available and readable during the entire interim storage period and possibly beyond. Since the storage time could span more than one human generation, transfer of information from one generation to the next is important.

5. CHARACTERIZATION FOR STORAGE

Paramount to the safe and practical management of RRSNF is a comprehensive understanding of the characteristics of the fuel from the time that the fuel is removed from the reactor until long after the fuel is disposed. This section provides a discussion of data that may be required to support that understanding. A discussion of baseline characterization needs and tools available for spent fuel characterization is provided.

5.1. FUEL DATA NEEDS

The initial composition and physical geometry of research reactor nuclear fuel elements vary widely. While the majority consist of a meat containing the fissile material encased in aluminium clad, in some cases stainless steel, Inconel and zircaloy have been used as cladding material.

The meat can be formed by many composites, like uranium oxide, uranium silicide, uranium carbide particles in an aluminium matrix, pure metallic uranium, aluminium–uranium alloy, uranium–molybdenum alloy, or an alloy of uranium zirconium hydride (UZrH). Powder metallurgy techniques are used in the manufacture of the fuel plate meats, making briquettes using ceramic or metallic composites. In general, the briquette is made with powdered nuclear material and pure aluminium powder, which is the structural material matrix of the briquette.

Traditionally, the enrichment of research reactor fresh fuel ranged from 20–90%. However, since 1978 many reactors have been converted from high enriched uranium (HEU) to low enriched uranium (LEU) fuel; that is, to enrichments below 20%. Some research reactors, which were especially designed to operate with a high density of ^{235}U per cubic centimetre, require the development of a high density LEU fuel prior to converting it to an enrichment below 20%. Regardless of the enrichment, the burnup of the spent fuel ranges from 30–70%, and after irradiation, the meat of the spent fuel consists of a mixture of fission products and the remaining uranium composite or alloy.

Considering that, in most cases, the purpose of a research reactor is to test new fuels and/or materials, the geometry of a research reactor fuel element also presents a large variety of shapes and dimensions [75–77], as can be seen in Fig. 11.

The most common type of fuel element used in typical western research reactors is the materials testing reactor (MTR) fuel element, shown in Fig. 11 (E). It is an assembled set of regularly spaced aluminium fuel plates, forming a fuel assembly in which the channels between two consecutive plates allow a stream of water that serves as coolant and moderator to the high energy neutrons produced in the nuclear reaction. The fuel plates can be flat or curved, depending on the adopted design, and have a meat containing the fissile material, entirely covered with aluminium. They are manufactured by adopting the traditional assembling technique of inserting a fuel briquette into a frame covered by aluminium plates, which are subsequently welded and submitted to a rolling process. This technique is known internationally under the name 'picture-frame technique'.

Different than the plate type fuel elements used in Western type research reactors, the most common type of fuel elements used in Russian type research reactors is a seamless tubular design, where the outlying layers are the cladding and the middle layer is a fuel composition, similar to the 'meat' used in

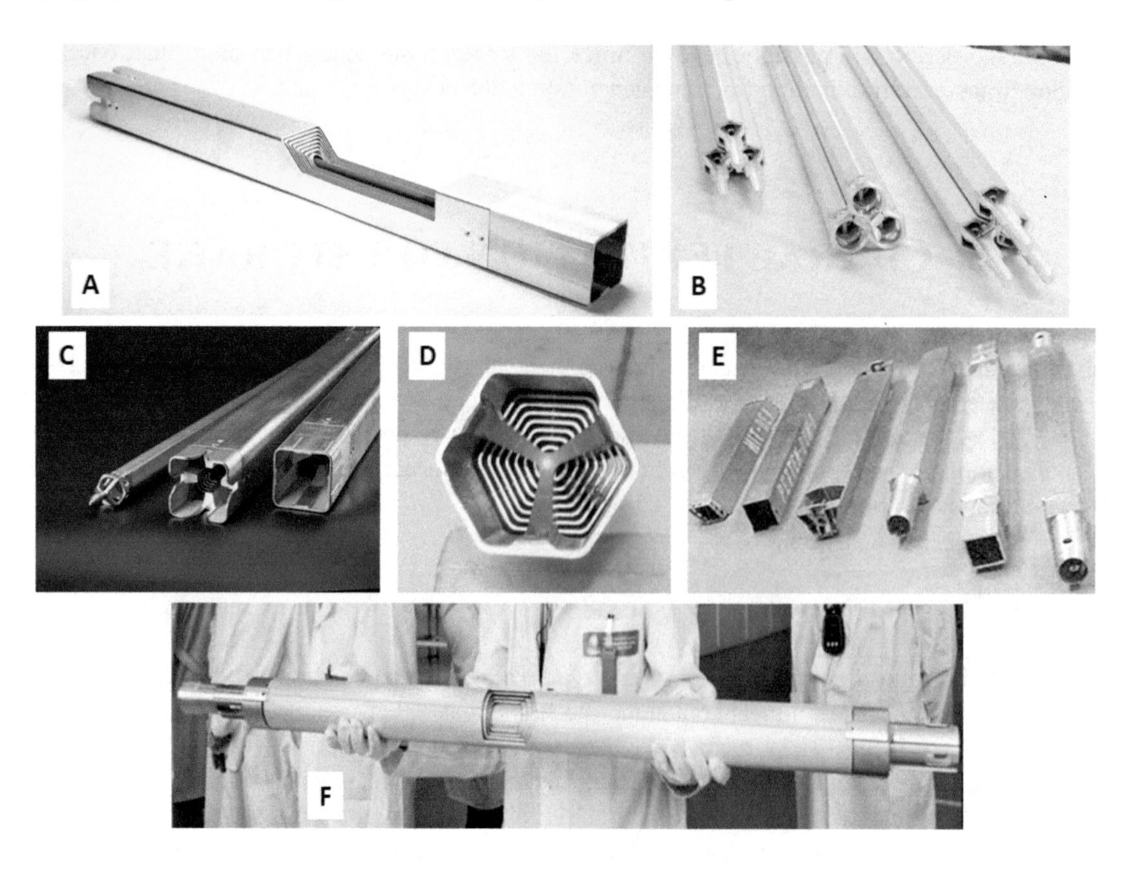

FIG. 11. Typical nuclear research reactor fuel elements from the Russian Federation (A)–(D) and the USA (E)–(F) (courtesy of NCCP [75–77], BWXT, and General Atomics).

MTR fuel elements. Final geometry is obtained by using a hot coextrusion process. By changing the tube profile, wall thickness and the active layer composition, it is possible to meet the requirements to fuel most Russian type research reactors [78].

Regardless of the type of reactor where it will be used, no two fresh fuel elements are identical. Acceptable variations of the manufacturing process are normal, resulting in slightly different fuel elements. These differences are measurable and need to be recorded to clearly identify the specific fuel element.

In addition to the manufacturing data, each research reactor fuel element has a different irradiation history. After being removed from the reactor core, the length of time it is maintained in water to achieve a specific decay is set for each fuel element depending on its specific characteristics.

When the fuel element is ready to go to interim storage, there is an extensive data history that is used to store it safely and effectively with all other fuel elements. These data support evaluations of criticality safety, shielding and thermal performance of the spent fuel assemblies during the interim storage period and beyond. Additional information on the operational history, physical condition of the spent fuel assembly, and any parameter affecting its physical condition, such as corrosion, is necessary to assure handleability and the structural integrity of the fuel assemblies during storage.

5.1.1. Design data

5.1.1.1. Assembly design details

In order to provide reliable post-discharge characteristics of a particular spent fuel assembly, it is necessary to have the fuel assembly design details. The minimum requisite information for accurate estimates of spent fuel characteristics include total fissile loading, initial enrichment, fuel meat composition and form, and composition of the cladding material, in addition to fuel meat and cladding dimensions, number of plates or rods per assembly, and the centre-to-centre spacing or pitch of the individual fuel plates or rods within the assembly. Additional detail on non-fuel components within a fuel assembly, including non-fuel bearing plates and other fuel hardware, can provide additional accuracy on radiation source term assessments of post-discharge fuel assemblies, especially for materials prone to neutron activation during service (e.g. stainless steel components). Typically, the non-fuel components may be ignored from a neutronics perspective, due to the relatively short half-life exhibited by most activation products. However, these materials may prove important from a fuel degradation perspective due to potential galvanic corrosion phenomena where dissimilar materials are in contact.

The corrosion degradation resistance of fuel assemblies is impacted by the construction materials, manufacturing methods (joining technology, fuel clad bonding technology, etc.), assembly geometry, and service and handling induced changes. Therefore, details on materials and methods used in producing nuclear fuel assemblies are essential in order to understand the condition of the fuel decades after it has been removed from the reactor core. This needs to be recorded and maintained throughout the life cycle of the assemblies. In addition, observed changes in the fuel that are induced by reactor service and handling also have to be noted and those notes maintained.

5.1.1.2. Physical form

Information on the physical form of the fuel being stored is important for assessment of the potential risk to safety and impact on facility operations due to the storage of SNF. This information will likewise be required for future transport and disposal of the spent fuel. The information addressed needs to include whether the fuel material is in the form of sintered oxide pellets (declad), loose oxide powder, fuel particle dispersions or metallic alloys.

5.1.1.3. *Design enrichment*

The concentration of fissile isotopes within the fuel assembly relative to the quantity of potential absorber isotopes is a fundamental fuel characteristic relative to criticality safety. Fuel enrichment provides the most important metric with regard to criticality safety requirements for the life cycle of a fuel assembly. Often, due to uncertainty in fuel depletion calculations, and in order to provide a safety margin in criticality calculations, the original, as-fabricated enrichment is used for criticality calculations. The use of as-fabricated fuel assembly enrichment data in criticality calculations typically provide for the most conservative results and thus in criticality safety assessments. Criticality safety assessments based on initial enrichment typically result in the largest criticality safety margins for spent fuel storage.

5.1.1.4. *Reactor design details*

Accurate assessments of as-discharged SNF assemblies require information on several key reactor design features. The minimum set of data includes coolant and moderator composition and number, location and spacing of fuel and non-fuel components within the reactor core. Details on design service temperatures of the moderator and coolant and moderator density will reduce reliance on modelling assumptions and, consequently, reduce uncertainty in model results. Reactor lattice details related to assembly spacing or pitch, and location, composition and use of neutron population control features will likewise reduce modelling uncertainties related to input assumptions.

5.1.2. Post-discharge nuclear data

Safe handling and storage of SNF requires an understanding of composition of the fissile, fertile and radioactive isotopes contained within the spent fuel assembly. Post-discharge nuclear data that supports this understanding is required to manage potential risks associated with inadvertent criticality events, external radiation exposure and thermally induced physical changes to the spent fuel assembly. These nuclear data are discussed in Sections 5.1.2.1 to 5.1.2.5.

5.1.2.1. *Enrichment*

In general, post-discharge enrichment is determined by calculations using well established reactor physics codes, such as those discussed in Section 5.3.2. Typically, the same code that was used to follow the history of the fuel while it was in the reactor core is used to calculate post-discharge enrichment. However, due to uncertainty in fuel depletion calculations, the original, as-fabricated enrichment is often used for criticality calculations, to provide a safety margin. Although the use of as-fabricated fuel assembly enrichment in criticality calculations is typically conservative, there are some fuel and target assemblies that become more reactive, from a nuclear criticality perspective, with continued reactor exposure that will necessitate special consideration. These assemblies will require the most comprehensive set of data regarding exposure history to ensure criticality safety during fuel handling and storage. In addition, accurate knowledge of the quantities of fissile isotopes, relative to absorber isotopes, within the fuel assemblies after discharge may reduce reliance on criticality safety margins during spent fuel storage.

5.1.2.2. *Isotopic composition*

The isotopic composition of spent fuel affects several aspects of the safe storage, handling and disposal of the fuel assemblies. At the end of life, SNF contains a mixture of stable and radioactive fission products, actinides and activation products. Knowledge of the quantity and distribution of each species within the SNF with time is paramount for safe and effective storage, handling and eventual disposition. The fission products represent the predominant source of radiation from the fuel assembly. In addition, fission product decay drives the thermal power of a spent fuel assembly for decades after assembly discharge. Some fission products are very soluble in water (^{137}Cs, most notably), and if the fuel element develops some leaks, these products become a major source of contamination in spent fuel storage pools. Actinides generated during reactor exposure include predominantly long lived heavy isotopes. Actinides tend to contribute little to the external dose; however, the concentration of fissile actinides generated during exposure has to be considered in criticality safety analyses. Actinide decay heat tends to contribute little to the overall heat generated in the spent fuel assemblies during interim storage compared to that generated by fission product decay. As in the case of post-discharge enrichment, in general, post-discharge isotopic composition is also determined by calculations based on well established reactor physics code, such as those discussed in Section 5.3.2 [79–81].

5.1.2.3. Burnup

Fuel assembly burnup is indicative of how much of the assembly fissile material underwent fission during reactor operation, and provides a metric of the amount of energy generated from the assembly during reactor service. Burnup data allow estimation of the number of fissions and resulting fission product inventory resulting from irradiation in the reactor core for a specific assembly. Often, assembly burnup is estimated based on knowledge of reactor power history and the number of assemblies within the reactor during operation. This type of burnup information represents the minimum set of required data on fuel exposure to facilitate estimates of end of life fuel composition. The estimates based on these limited data may have significant uncertainty relative to estimates based on the more comprehensive set of exposure data discussed in Section 5.1.2.4, but may be sufficient where a significant safety margin is acceptable. Furthermore, the uncertainty in calculations based on average core burnup data is significantly lower for reactors with a relatively flat axial and radial flux profile and with relatively consistent power history.

5.1.2.4. Exposure history and records

As mentioned previously, assembly burnup is often estimated based on knowledge of reactor power history and the number of assemblies within the reactor during operation. However, use of average reactor power over a given period of time or even using time-dependant, total reactor operating power to estimate the burnup of a specific assembly introduces uncertainty in calculations performed to estimate end of life isotopic data for the assembly. The minimum set of requisite exposure history information to facilitate an assessment of discharged spent fuel includes: initial service date, average reactor power during service, average coolant temperature, and assembly discharge date. Additional exposure history data that can significantly reduce uncertainty in the characterization of SNF assemblies include a detailed time-dependent record of the reactor power operating history, assembly irradiation location, and temperatures at different times and locations within the reactor during service.

5.1.2.5. Thermal load and output

Understanding the thermal load generated within SNF by radioactive decay is important to the design of safe interim storage configurations. This is particularly true for dry storage configurations. The majority of RRSNF is composed of low materials with low melting points such as aluminium or aluminium alloys. Therefore, it is important to preclude fuel assembly creep during interim storage, which will impact the handleability, retrievability and structural integrity of the fuel. For wet storage configurations, the heat is removed by convective transfer of the heat from the fuel assemblies to the bulk basin water. However, dry storage configurations result in significantly reduced convective heat transfer of decay heat from the fuel assemblies, potentially leading to higher temperatures in both the fuel and cladding. If the storage environment is humid, the mechanisms for corrosion degradation of SNF assemblies can be enhanced, especially at higher temperatures. Accurate values of the thermal heat load of spent fuel assemblies is essential to designing dry storage configurations with sufficient heat rejection capabilities to preclude deformation of the fuel assemblies due to creep. In general, the methodology [80] used to estimate the decay heat produced by spent fuel after it is removed from the reactor core is based on an empirical equation attributed to Way and Wigner. Accordingly, for a fuel assembly irradiated continuously during a time, t_i, at a constant fuel assembly power, P_0, the heat load power $P(t_d)$ per assembly at time after irradiation, t_d, is given by:

$$P(t_d) = 6.85 \times 10^{-3} \times P_0 \times (t_d^{-0.2} \quad (t_i + t_d)^{-0.2})$$

All power units are expressed in watts and time units are expressed in days. Fuel assembly decay heat loads calculated with the Way–Wigner equation are expected to be conservative, and within a factor of two or less of measured heat loads. The thermal heat load of a fuel assembly is independent of the fuel assembly type [80].

5.1.3. Physical condition

The resistance of SNF assemblies to degradation is highly impacted by physical changes that may be caused by reactor service and handling. During storage, corrosion induced changes to the spent fuel may exacerbate existing corrosion mechanisms. Therefore, an accurate and ongoing understanding of the physical condition of the fuel, and whether the cladding is present, intact, degraded or absent, is imperative to a safe and effective management programme for SNF storage.

5.1.3.1. Geometry

Details on the SNF assembly geometry are important inputs to post-discharge assessments required for effective fuel management. Given that details are available for the as-fabricated assembly geometry, records have to be maintained for physical changes to the assembly resulting from reactor service (e.g. swelling or warping) or due to any intentional reconfiguration or alteration of the fuel assembly intended for storage. Intentional reconfiguration would include decladding, cropping or reconstitution of fuel assemblies prior to spent fuel storage. Additionally, detailed information on the physical characteristics of spent fuel assembly pieces and parts generated as a result of destructive post-irradiation examination have to be documented and maintained until those incomplete fuel assemblies are disposed of.

5.1.3.2. Extent of degradation

The extent to which SNF degrades can potentially impact both safety and facility operations. The ability to safely store SNF requires an ability to confine the radioactive constituents of the fuel. This is more challenging for fuel with high levels of degradation. Cladding degradation can lead to exposure of fuel material to the storage system environment, leading to higher likelihood of loss of confinement, which could also impact the criticality safety of the storage system and the radiation management of the facility, as well as increase the potential for release of radioactive material. Normal operations at the facility might be impacted by confinement compromise, as indicated by increased operational requirements of water filtration and deionization systems for wet storage systems and increased ventilation filtration requirements for dry storage systems. Increased filtration leads to increases in waste generation and handling requirements. Because of the potential impacts to facility safety and operations, it is important to maintain a current understanding of stored fuel degradation. This information is generally gathered through periodic inspection of the fuel and fuel surrogates as part of a comprehensive strategy for management of spent fuel storage systems. Data gathered about fuel degradation need to be documented and maintained throughout the life cycle of the fuel until its disposal.

5.1.3.3. *Structural and cladding integrity*

In a management programme for SNF storage, it is paramount to maintain the ability to safely handle the spent fuel until its disposal. Unmitigated localized degradation of spent fuel can compromise the structural integrity of the fuel assembly or integral lifting hardware, jeopardizing the ability to readily handle the fuel assembly. In addition, persistent and unmitigated localized degradation attack at joints of the assembly can compromise the ability of the fuel to maintain its original configuration and, hence, potentially impact safety analysis assumptions. Because of the potential impacts to facility safety and operations, it is important to ensure the structural integrity of the SNF and to identify and address any localized fuel degradation. This is typically accomplished through the development and implementation of a management strategy for the SNF storage systems [20]. Data collected regarding the potential for impaired handling ability of individual fuel elements have to be documented and maintained throughout the life cycle of the fuel.

5.2. BASELINE CHARACTERIZATION NEEDS

In order to assess changes in the spent fuel assembly condition during interim storage, it is essential to have a comprehensive understanding of the condition of the individual assemblies prior to emplacing for storage. In addition to the fuel data needs discussed previously, an assessment has to be conducted and documented for each SNF assembly prior to emplacement into any storage system. The records from this initial assessment provide the baseline against which subsequent assessments may be compared to determine the impact of storage conditions on the fuel assembly condition over time. The combination of fabrication, operation and storage history data, along with records of the baseline assessment prior to emplacement, provide the foundation upon which the basis may be developed to demonstrate continued safe storage of the spent fuel assemblies being introduced into the new storage environment or configuration.

5.2.1. Baseline characterization for wet storage

The baseline characterization of SNF assemblies that should be performed prior to introduction of a spent fuel assembly into a wet basin storage facility includes the design and nuclear data discussed previously along with the records related to the exposure history and documented physical condition of the fuel assembly. In addition, a comprehensive visual inspection and a sipping test has to be conducted and documented on the individual assemblies, as discussed in subsequent sections of this publication. The design and nuclear data along with the historical records on the exposure history will be used to complete assessments of the assembly composition, the radiation source term and the potential for criticality. In addition, predictive models may be used to demonstrate integrity of the fuel during the intended storage period. These data and documented results from the safety assessments provide the complete set of baseline characterization data required to demonstrate continued safe and effective spent fuel wet storage.

5.2.2. Baseline characterization for dry storage

The baseline characterization of the SNF assemblies intended for a dry storage facility include all those components required for a wet storage system as well as details on the as-loaded dry storage configuration. Details on the canister materials and geometry, the basket materials and basket configuration, the backfill gas, and the quantity of residual water after drying are recorded. These data will provide input parameters for safety assessments regarding criticality, radiation shielding and thermal performance. In addition, predictive models may be used to demonstrate integrity of the fuel during the intended storage period. The data and documented results from the safety assessments provide the complete set of baseline characterization data required for the demonstration of continued safe and effective spent fuel dry storage.

5.2.3. Baseline characterization for overpacking

The baseline data required to support overpacking of SNF is consistent with that data collected for wet and for dry storage options. Any additional information regarding changes to the fuel or fuel containers after original emplacement and prior to overpacking needs to be combined with previous baseline data to develop a new baseline dataset for the fuel. These data would include changes to the original geometry of the fuel and canister due to handling activities or due to corrosion degradation. For example, the occurrence of a breach of containment of a storage package and subsequent loss of the inert backfill gas or the ingress of water needs to be recorded as part of the baseline characterization package of data for the new overpack.

5.3. CHARACTERIZATION TOOLS

There are two general categories of characterization tools available for the assessment of the condition of SNF assemblies. The first category includes tools that are used to directly interrogate spent nuclear assemblies and fuel storage canisters. The second category includes modelling and simulation software used for the estimation of the physical attributes of the fuel and canister. Typically, data developed from the physical examinations are used to validate the predictive models developed. In general, validated modelling and simulation tools provide significantly better resolution of spent fuel attributes than do tools available for physical measurements with relatively significant cost savings. However, reliance on modelling and simulation tools without the support of physical measurements needs to be avoided, when possible, to minimize requisite safety margins and uncertainty.

5.3.1. Physical measurement tools

Physical measurements of key attributes of spent fuel assemblies are crucial to support initial and continued fuel storage in either wet or dry configurations. There is a variety of tools and equipment available to collect data and information required to determine the condition of SNF assemblies, and to support a demonstration of the viability of continued safe and effective storage [82]. These tools may be grouped together as supporting either non-destructive or destructive examinations of the fuel assembly condition. Some tools that fit into these two broad groups are discussed in Sections 5.3.1.1 and 5.3.1.2.

5.3.1.1. Tools for non-destructive examination

There are several tools and approaches available for examination of SNF assemblies that do not require dismantling or extraction of materials, and thus enable NDE. The most commonly used approach is the visual inspection. It is also the most versatile, useful and cost-effective method of interrogation for SNF assemblies. Qualified, radiation resistant video and still cameras that meet ASME Section V, Article 9 regarding visual examination [28] will last longer than standard video cameras when used for visual inspections. Measuring equipment, including rulers and micrometers, can be employed to assist in accurate profilometry characterization.

In addition to visual inspection methods, balances can be used to weigh the assembly. Ultrasonic testing and, to a lesser extent, eddy current testing equipment can be used to identify pits or cracks in fuel cladding. Gamma spectroscopy can be used to gain some understanding of the fission product composition of the fuel based on the spectrum of emitted gammas. Nuclear magnetic resonance, scanning neutron radiography and gamma radiography techniques may be used to determine water content in the void space of some fuel designs.

5.3.1.2. *Tools for destructive examination*

There are several tools and analytical methods that are suitable for laboratory examination of samples that have been harvested from SNF assemblies. The harvested samples can be of varying size, mass and shape and may have been extracted from any area of the fuel assembly of interest. Methods available for the characterization of fuel samples include optical microscopy [83], scanning electron microscopy [84], transmission electron microscopy [84], X ray diffraction and X ray fluorescence [85]. In addition, dissolved solutions of spent fuel samples can be analysed for composition using thermal ionization mass spectrometry, inductively coupled plasma mass spectrometry, spark source mass spectrometry, alpha spectrometry and gamma spectrometry.

5.3.2. Computer simulation tools

Existing computer simulation tools are mainly used to simulate isotopic transmutation and decay within the nuclear fuel element of an operating nuclear reactor. This provides an accurate and defensible means of estimating the composition, radiation source term and thermal load of the SNF assemblies over time during reactor service and after discharge. At present, numerous commercial code packages exist that are being widely used to perform integrated lattice core physics and isotope depletion simulations. The list of available, non-proprietary code packages includes SCALE, MCNPX, Monteburns, DARWIN, CASMO and BGCore [86–88]. In addition, codes like ORIGEN and CINDER are available for calculating material depletion and decay given user-defined neutron flux information [89]. Most of the aforementioned codes and code packages are being actively developed and maintained to increase capability, reduce uncertainty and update data libraries as new or revised nuclear data. More powerful and efficient calculational methods are being developed as well. There are numerous other, proprietary codes and code packages that are routinely used in the nuclear industry. One limitation of computer simulation tools is that they can be used only to predict the isotopic composition of the fuel assembly. They cannot be used to determine any degradation, in terms of physical damage, such as corrosion suffered by the fuel assembly while in the reactor core.

5.4. FUEL CONDITION ASSESSMENT

The following Sections, 5.4.1 to 5.4.3, provide brief information about assessing the physical condition of SNF prior to and during interim storage. Included are methods suggested for conducting fuel assessments and a general guide to identifying and recording indications of fuel degradation during storage. Key elements of the fuel surveillance programme include periodic visual inspection of individual fuel assemblies, or a representative subset of the spent fuel assembly inventory, sip testing of breached fuel assemblies, and non-destructive testing techniques for assessing the integrity of sealed packages. These elements are discussed in further detail below.

5.4.1. Visual inspections

A visual inspection programme includes visual inspection of the individual fuel assemblies, where feasible, or a representative subset of the total spent fuel inventory. To facilitate effective visual inspection, an inspection station includes an underwater inspection table to be used for the methodical inspection of single spent fuel assemblies. Figure 12 shows a typical and simple arrangement that can be used for visual inspection of RRSNF. The inspection station employs implements capable of rotating fuel assemblies on the inspection table during the inspection relative to the position of video and still image recording equipment. The table has graduated measuring implements that are fixed to the table that are used to determine image magnification level, where zoom camera equipment is used. The camera lens is maintained at a known and fixed distance from the surface of the table throughout the inspection for consistency in image magnification. Otherwise,

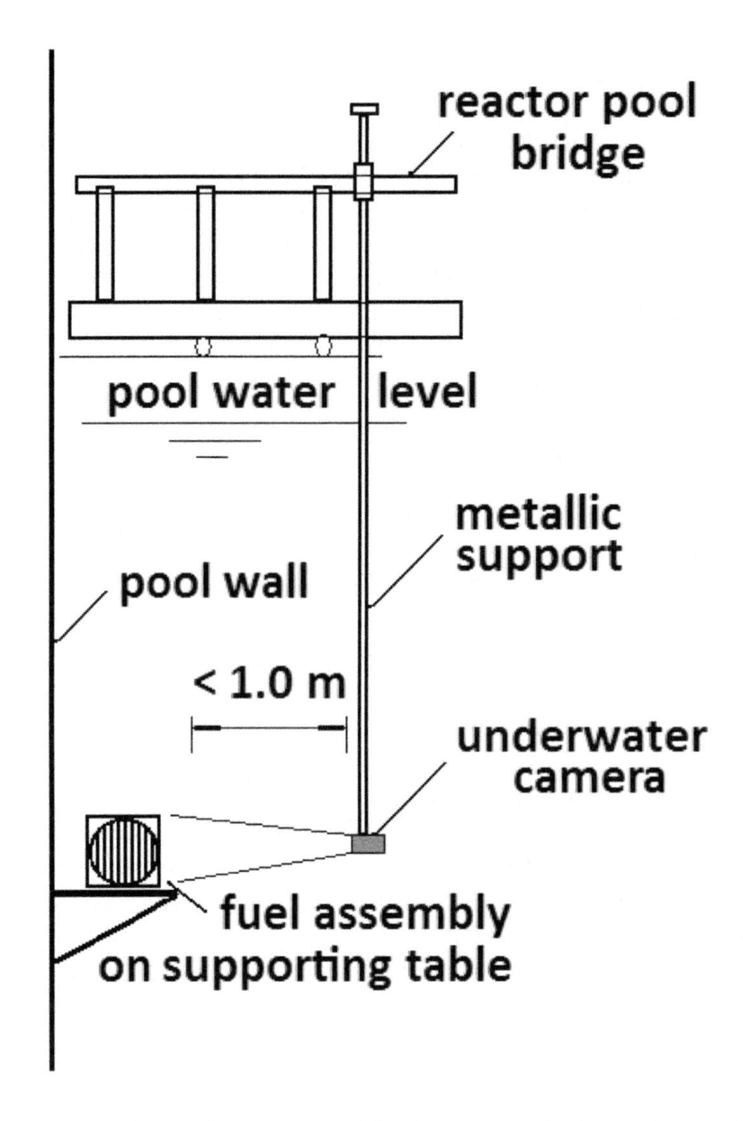

FIG. 12. Typical arrangement for visual inspection of RRSNF (courtesy of the Savannah River National Laboratory).

implements need to be available to provide an accurate measure of the distance between the fuel surface and the camera.

An effort has to be made to develop a consistent method of identifying each of the distinct sides of the spent fuel assembly of interest, to facilitate comparison of inspection findings between multiple fuel inspections of the same assembly. The presence of a stamped assembly identification number on one assembly surface can greatly assist in maintaining consistency in labelling sides of assemblies. Likewise, fuel with concave plates is better suited for consistent identification of the specific side being viewed; however, the orientation of the assembly from end to end can become confused if other indicators are not available. It is suggested that a chloride-free permanent marker may be used to mark one end of one side plate of an assembly in order to maintain consistency of identifying particular sides of assemblies where other indications (e.g. assembly ID numbers) are not present. Once the methodology has been established, the spent fuel can be visually inspected, to detect any degradation that can occur during the lifetime of the fuel assembly.

Radiation resistant endoscopes, already available in the international market, can be used for inspection of the internal parts of tubular fuel elements from the Russian Federation, and the internal plates of MTR fuel elements.

5.4.2. Sip testing

SNF assemblies that are suspected of having breached cladding can be subject to testing to determine the rate of release of radioactive isotopes to basin water. The most common test used for this determination is the sip test. To perform the sip test, a single assembly is isolated from bulk basin water, and inserted in the sip testing apparatus. The water within the sip testing apparatus — typically a pipe that is sealed on the lower end and located within the basin with the upper end above the level of the pool water — is flushed with clean, demineralized water. Immediately after flushing operations are complete, the water is mixed by passing compressed air through the volume, and a sample of the water is taken and analysed to measure the baseline ^{137}Cs activity of the water. The flooded system containing the fuel assembly remains isolated from bulk basin water for a predetermined time, typically four hours. The water within the system is then mixed again, using compressed air, and a second sample is taken and analysed for activity. Comparison of the samples allows for the determination of whether the fuel assembly has exposed fuel meat as well as the leak rate for those assemblies determined to be leaking. In addition, the leak rate of the fuel assembly can be used with the known degradation rate of exposed fuel meat and specific activity of ^{137}Cs within the fuel meat to estimate the surface area of fuel meat that is exposed through the breaches in its cladding.

5.4.3. Data needed to characterize physical condition

This section establishes a minimal set of the data that need to be maintained in the characterization of RRSNF assemblies. The data discussed, along with design and nuclear data plus photographic records, provide the foundation on which analyses and assessments are developed to demonstrate the viability for a continued safe and effective interim storage of the SNF.

The as-built geometry of the SNF assembly along with any alterations to the physical geometry has to be recorded. The alterations include intentional alterations, such as fuel cropping or fuel sectioning activities, and those alterations that are induced by handling, reactor service, or degradation phenomena.

The physical location of welds in the spent fuel assembly has to be recorded along with an indication of the welding technique used in performing the joining and any weld filler materials used.

Cladding flaws including pits, blisters, or cracks have to be recorded along with the relative location and size of the individual defects. Flaws that have been determined to penetrate the cladding to expose fuel meat need to be noted. The estimated surface area of spent fuel meat likewise needs to be noted. In addition, cracks discovered that could compromise the structural integrity of the fuel assembly have to be recorded.

Corrosion product buildup between adjacent fuel plates or localized attack at the interface between a fuel plate and adjacent side plate has to be recorded. An attempt should be made to note the extent of corrosion attack to facilitate monitoring of the progression of the degradation phenomena for subsequent inspections of the subject fuel assembly.

The primary concern with creep induced alterations of spent fuel assemblies relates to retrievability of the fuel assemblies. To this end, creep induced changes that could impact the ability to emplace or retrieve the fuel assembly or its structural integrity need to be recorded. An example of this sort of physical change would be the potential warping of a fuel assembly to the extent that it becomes bound in the storage location.

The rate at which breached SNF releases its radioactive contents has to be accommodated by the deionization system of the storage facility. Therefore, known leak rates of a fuel assembly, typically determined by sip testing, have to be recorded and maintained as part of the characterization package for SNF assemblies.

6. SAFETY ISSUES FOR RESEARCH REACTOR SPENT FUEL STORAGE

6.1. SAFETY GUIDE ON STORAGE OF SPENT NUCLEAR FUEL

The management of spent fuel from NPPs and research reactors is an important concern relating to the peaceful use of nuclear energy. Projects for the implementation of deep geological disposal are being delayed, and many States have yet to decide on the final destination of spent fuel generated on their territory. Consequently, long term storage is becoming a reality.

IAEA Safety Standards Series No. SSG-15 (Rev. 1), Storage of Spent Nuclear Fuel [20], was published in 2020. The scope of this Safety Guide includes the storage of spent fuel from water moderated reactors and can, with due consideration, also be applied to other fuel types, such as those from gas cooled reactors, research reactors and also to spent fuel assembly components and degraded or failed fuel that may be placed in canisters.

The objective of the Safety Guide is to provide up-to-date guidance and recommendations on the design, commissioning, safe operation and assessment of safety for different types of spent fuel storage facilities (wet and dry), considering different types of spent fuel from nuclear reactors, including research reactors, and different storage periods, including storage going beyond the original design life of the storage facility.

The Safety Guide presents guidance and recommendations on how to fulfil the safety requirements established in the following IAEA Safety Standards Series publications: SSR-4, Safety of Nuclear Fuel Cycle Facilities [90], GSR Part 5, Predisposal Management of Radioactive Waste [91], GSR Part 4 (Rev. 1), Safety Assessment for Facilities and Activities [92], and GSR Part 2, Leadership and Management for Safety [93].

Annex I of SSG-15 (Rev. 1) [20] defines short term storage as lasting up to approximately 50 years and long term storage beyond approximately 50 years with a defined end point (reprocessing or disposal). SSG-15 (Rev. 1) [20] also states that long term storage is not expected to last more than approximately 100 years. This time span is judged to be enough time to determine future steps in spent fuel management. Continued generation and storage of spent fuel without full commitment to a clearly defined end point is not a sustainable policy. SSG-15 (Rev. 1) [20] states in Paragraph 1.6 that "storage cannot be considered the ultimate solution for the management of spent fuel, which requires a defined end point such as reprocessing or disposal in order to ensure safety."

6.2. REGULATORY REQUIREMENTS

The national regulatory body, as established by the government, is responsible for developing basic safety criteria, establishing national regulations, and granting authorization to an operating organization or to an individual, to operate the facility. However, it is the organization responsible for the facility, also known as the licensee, who has the prime responsibility for the safety of the activities that give rise to radiation risks at the specific facility [2, 94].

Generally, the licensee retains the primary responsibility for safe management of the spent fuel. Other groups, such as designers, manufacturers, constructors, employers, contractors, consignors and carriers, also have legal, professional and functional responsibilities with regard to safety, but usually for specific activities and for a limited time.

It is important to clarify that the responsibility of the government goes beyond the responsibility of the licensee which operates a facility with nuclear material. This is because of the long period needed to store the spent fuel, which easily goes beyond the lifetime of the facility. The long period for which the spent fuel is isolated (or its waste in case of reprocessing), exceeds the present generation and increases the responsibility of the government as the ultimate legatee of the spent fuel or its waste, in case of reprocessing. It invokes organizational, financial and economic issues as well as having safety and technological consequences, justifying why the government has the responsibility to define the national policy on nuclear waste, to carry out policy oversight and regulation, to define the processes for funding, siting and environmental assessment of the facilities, and possibly implements them. These responsibilities are similar to other industrial activities, however, the long time scale associated with spent fuel management adds some unique features [95].

A facility for SNF storage is a nuclear facility, so all general requirements for nuclear facilities would be applicable as well. IAEA Safety Standards Series No. SSR-1, Site Evaluation for Nuclear Installations [96], establishes safety requirements for the siting process for all types of nuclear installation, including facilities for SNF storage.

6.3. COMMISSIONING

The process of commissioning of a RRSNF storage facility will follow a logical progression of steps intended to demonstrate that all equipment and SSCs of the facility function properly and as expected. During commissioning, the operating procedures for normal operation, anticipated operational occurrences and accident conditions are verified and the readiness of staff to operate the storage facility is demonstrated. Facility commissioning needs to be considered as part of the facility design process, and all commissioning plans need to be reviewed by a diverse group of technical and regulatory specialists. Additional recommendations on commissioning of spent fuel storage facilities can be found in SSG-15 (Rev. 1) [20].

6.4. PHYSICAL PROTECTION

Due to the potential consequences of the theft and misuse of nuclear material or sabotage against the nuclear installation, security is an important complement to safety for any nuclear facility, and stringent measures are required for their physical protection.

The prevention of unauthorized access or material removal requires the implementation of stringent security and access controls. Such controls need to be compatible with the safety measures applied at the facility [20]. Recommendations and guidelines on arrangements for physical protection at nuclear facilities are provided in Refs [97, 98].

According to the provisions of the Convention on the Physical Protection of Nuclear Material [97] and the Amendment thereto [99], the responsibility for the establishment, implementation and maintenance of a physical protection regime within a State rests entirely with that State. The State's physical protection regime is intended for all nuclear material in use, in storage and during transport, and for all nuclear facilities. To meet its obligation, the State needs to ensure that the nuclear material and nuclear facilities are protected against unauthorized removal and sabotage. This includes a regular revision and update of the physical protection regime to reflect changes in the threat, advances made in physical protection approaches, systems and technology, and the eventual introduction of new types of nuclear material and nuclear facilities.

Site selection and design for a RRSNF storage facility need to take physical protection into account as early as possible. Consideration needs to be given also to the interface between safety, security (including physical protection) and nuclear material accountancy in the design and operation of the facility [20].

6.5. SAFEGUARDS

Accounting for and control of nuclear material is done in accordance with safeguards agreements between the Member State and the IAEA, with the objective of timely detection of the diversion of nuclear material for non-declared purposes and deterrence of such diversion by early detection. The IAEA, through its safeguards role, is responsible for providing independent, international verification that governments are abiding by their commitments with respect to the Treaty on the Non-Proliferation of Nuclear Weapons [100]. According to the treaty each non-nuclear-weapon State that is party to it undertakes the obligation to accept safeguards in an agreement negotiated and concluded with the IAEA. This agreement is for the exclusive purpose of verifying the fulfilment of obligations assumed under the treaty to prevent the diversion of nuclear energy from peaceful uses to nuclear weapons or other nuclear explosive devices.

The IAEA's safeguards system consists of several, interrelated elements: (i) the IAEA's statutory authority to establish and administer safeguards; (ii) the rights and obligations assumed in comprehensive safeguards agreements and additional protocols; and (iii) the technical measures implemented pursuant to those agreements, which in some cases can be performed by an agency endorsed by the IAEA, such as the European Atomic Energy Community (called Euratom) or the Brazilian–Argentine Agency for Accounting and Control of Nuclear Materials (called ABACC). Together, these elements enable the IAEA to independently verify the declarations made by Member States about their nuclear material and activities. The nature and scope of such declarations — and of the measures implemented to verify them — stem from the type of safeguards agreement that a Member State has in force with the IAEA. Under such agreements, the IAEA is also responsible for verifying the absence of possible undeclared material and activities. For the IAEA to be able to do so credibly, it is necessary that Member States have in force additional protocols to their comprehensive safeguards agreements, based on models approved by the IAEA.

As in the case of safety and security, it is the government's responsibility to ensure the implementation of ways to support safeguards activities. The IAEA safeguards system is based primarily on the use of materials accountancy as a safeguards measure, with containment and surveillance as complementary measures. In the context of storage operations, arrangements are made to ensure that the facility operator is aware at all times of the location and quantities of nuclear materials in storage and to provide the necessary reports as defined within the particular safeguards agreement between the Member State and the IAEA. An overview of IAEA safeguards verification methods for spent fuel in wet and dry storage is presented in Ref. [101].

6.6. EMERGENCY PLANNING

As part of the licensing process of a RRSNF storage facility, the operating organization has to develop an emergency plan, consistent with the requirements of the national regulatory body. SSG-15 (Rev. 1) [20] recommends that the potential radiological impacts of accidents be assessed and provisions be made to ensure that there is an effective capability to respond to all accidents. Further, IAEA Safety Standards Series No. SSG-47, Decommissioning of Nuclear Power Plants, Research Reactors and Other Nuclear Fuel Cycle Facilities [102], states in Paragraph 8.31:

> "The licensee is required to ensure that adequate resources, including personnel, equipment, means for communication, logistical support and emergency response facilities, are available and that procedures, coordination and organization are in place in accordance with the approved emergency plan. Personnel should be qualified, trained in emergency procedures and fit for duty, and consideration should be given to the need for the periodic review and updating of these procedures by means of regular exercises."

Inspections need to be performed regularly to ascertain whether equipment and other resources necessary in the event of an emergency are available and in working order [20]. General safety requirements and recommendations, respectively, related to emergency preparedness can be found in IAEA Safety Standards Series No. GSR Part 7, Preparedness and Response for a Nuclear or Radiological Emergency [103], and No. GS-G-2.1, Arrangements for Preparedness for a Nuclear or Radiological Emergency [104].

6.7. DECOMMISSIONING

While this publication is focused on spent fuel storage, facility decommissioning is an important step in the overall life cycle of any nuclear facility and needs to be considered throughout the lifetime of the facility, beginning with the initial planning and design of the facility, so it is briefly mentioned here. In IAEA Nuclear Energy Series No. NW-T-2.4, Cost Estimation for Research Reactor Decommissioning [105], "[t]he term 'decommissioning' is defined as the administrative and technical actions taken to allow the removal of some or all of the regulatory controls from a nuclear facility after its shutdown and the return of its site to a planned end state accepted by the national legislation."

In order to achieve the 'green field' state (i.e. release of the site for unrestricted use), actions and activities for long term protection of people and the environment have to be undertaken. These include decontamination, dismantling and removal of radioactive materials, demolition of the facility structures, and management of the waste and radioactive waste arisings. The time period necessary to achieve full decommissioning will depend on the type of installation, the radionuclide inventory, the chosen decommissioning strategy, the techniques employed and, in certain cases, the policy for radioactive waste management. However, it is fundamental that the timing of decommissioning, similarly to radioactive waste management, shall be such that it does not impose undue burdens on future generations in terms of both additional health and safety risks and financial requirements, as outlined in SF-1, Paragraph 3.29 [94].

7. LESSONS LEARNED FROM RESEARCH REACTOR SPENT FUEL STORAGE

This section provides several case studies to share details regarding situations where innovative solutions have been developed to address circumstances that have the potential to impact safe interim storage of RRSNF or to identify circumstances that may compromise fuel or cladding integrity during interim storage, and therefore impact safe storage operations. It is noted that the following case studies include figures, tables and sentiments that may have been presented previously in the current report, but have been reproduced herein to provide a more complete record of the case studies being described.

7.1. CENTRALIZED WET STORAGE AT SAVANNAH RIVER SITE

The L-Basin of SRS is part of the L-Area Complex, originally the home of the L-Reactor. The reactor started operation in 1954 and in 1988 was shut down for routine maintenance. However, in 1991, the USDOE decided not to restart it, defining its condition as permanent shutdown. The area of the building, also known as the disassembly area of the L structure, is an underwater storage facility that served as a cooling facility for L-Reactor's fuel during the facility's operational years.

In 1998, in order to avoid the cost of operating multiple facilities, USDOE decided to consolidate all the stored used fuel at SRS into the L-Basin and, since October 2003, all fuel previously stored in other locations of SRS has been moved to L-Basin for storage, leaving L-Basin as the only remaining SRS fuel receipt and storage facility.

Based on extensive analysis and monitoring, it is estimated that the L-Basin will maintain its structural stability until at least 2058 [106]. A basin lifetime study was initiated with the objective of obtaining empirical data to serve as a quantitative basis for the basin lifetime, to be reviewed against the structural analysis. In addition, the USDOE decided to start processing part of the SNF currently stored at the L-Basin, which liberates some of the storage capacity at the L-Basin, eliminating the need of additional storage during its lifetime [107].

7.1.1. L-Basin facility description

The Receiving Basin for Offsite Fuels (RBOF) was a wet storage basin (1.89 million litres) located also at SRS. Since the mid-1960s, the RBOF received and stored part of the RRSNF from off-site reactors. The RBOF had a good water quality: electrical conductivity levels between 1 and 3 μS/cm, maintained by continuously operated deionizers.

The L-Basin at the RBOF is a 12.78 million litre reinforced concrete structure, coated with a vinyl sealer paint [106]. Concrete walls are at least 76 cm thick, and the concrete floor is between 1.8 and 4.6 m thick. For more than 20 years, the water conductivity at the L-Basin was not optimized for corrosion control, with electrical conductivity levels maintained between 60 and 70 μS/cm by periodical use of portable deionizers. These deionizers were procured in the 1960s and for more than 20 years were shared by the several basins at SRS. In 1989, the fuel processing activities at SRS were stopped, which provoked the SNF to remain in wet storage at the basins that shared the deionizers.

The limited availability of the deionizers and the improper water conditions caused pitting and galvanic corrosion in the SNF stored at the L-Basin. In 1992–1993, the corrosion attack of the aluminium fuel and components was reported [108, 109]. Consequently, during the years 1994–1996, the USDOE decided to centralize the wet storage of unprocessed fuels, upgrading and refurbishing the L-Basin, and transferring all the SNF from the other basins and all new shipments of RRSNF from off-site reactors.

To ensure the accommodation of all the fuel in the L-Area during its operational lifetime, the Disassembly Basin Upgrade Project also considered actions to mitigate corrosion [18]. These started immediately with the installation of a new sand filter and home-designed portable deionizers in 1994. These helped improve the water conductivity and chloride concentration, and were followed by the installation of portable mixed bed deionization equipment in 1995, which further improved the water quality. In 1996, the equipment had to be repurposed and the original portable deionizers were again used at the L-Basin, helping achieve significant reductions of electrical conductivity and anion concentration.

7.1.1.1. Storage capacity

The L-Basin is divided into seven interconnected sections, which are 5.18 to 15.24 m deep [18]. The January 2013 L-Basin inventory included approximately 13 000 MTR equivalent fuel assemblies with aluminium cladding, 200 higher actinide targets and about 2000 non-aluminium based fuel assemblies with stainless steel or zircaloy cladding [110]. In the L-Basin, the fuel can be stored in several systems, depending on fuel design and fuel condition. There are vertical tube storage (VTS) racks, bucket racks, High Flux Isotope Reactor (HFIR) racks, oversized can racks and additional space for various non-standard customized configurations (see Fig. 49 in NP-T-5.2 [18]).

The VTS racks consist of a series of 3 × 10 or 4 × 10 modular racks that can be placed adjacent to one another. Between four and five MTR fuel assemblies are placed in an L-bundle, an aluminium cylindrical tube which is inserted vertically in the VTS rack, as shown in Fig. 13 (a). To allow water circulation, L-bundles have openings at ends, which also allow for water sampling, as shown in Fig. 13 (b). The L-Basin has 3650 VTS positions installed, with around 3000 filled with L-bundles as of 2018 [107].

The bucket storage area, shown in Fig. 14, is used to store fuel in a variety of containers (buckets) that are open at the top. The buckets are made of stainless steel and are used for fuels that are not amenable to be stored in L-bundles so they can be placed into VTS racks [111, 112].

The HFIR racks area is specifically organized to receive fuel elements from the HFIR, in which each fuel element is a complete core (Fig. 15). The HFIR racks area has a total of 120 positions designed to store HFIR fuel elements. In 2018 the inventory was 108 [107].

Degraded cladding or structural deformation of a fuel assembly, if significant, can compromise safety, waste and, eventually, accountability issues. Therefore, oversized storage can racks are used to store damaged or degraded fuel assemblies, separating them from the rest of the basin water. The damaged fuel and fuel pieces are placed underwater into small diameter cans, which are then grouped with others and placed in the oversized storage cans [113]. Figure 16 shows an oversized storage can being loaded with a tube containing a damaged fuel element, being closed and placed in the rack.

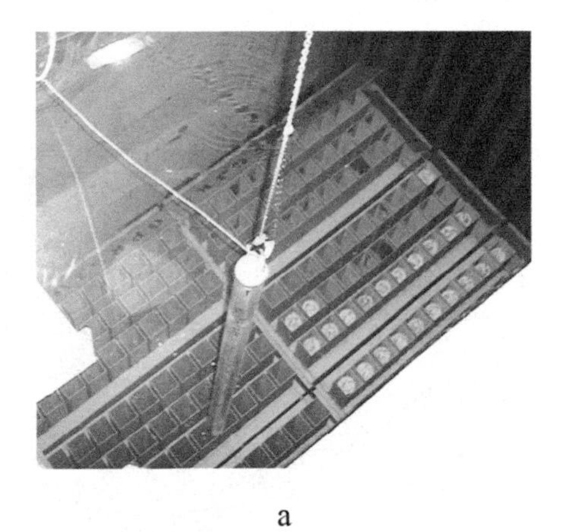

a

b

FIG. 13. (a) L-bundle being inserted into the installed 3 × 10 VTS racks and (b) pulling water sample from within bundle (courtesy of the Savannah River National Laboratory, USDOE [106, 110]).

a

b

FIG. 14. (a) Top view of bucket storage area of L-Basin showing (b) some stored components (courtesy of the Savannah River National Laboratory, USDOE [111, 112]).

FIG. 15. The HFIR fuel element (courtesy of the Savannah River National Laboratory, USDOE).

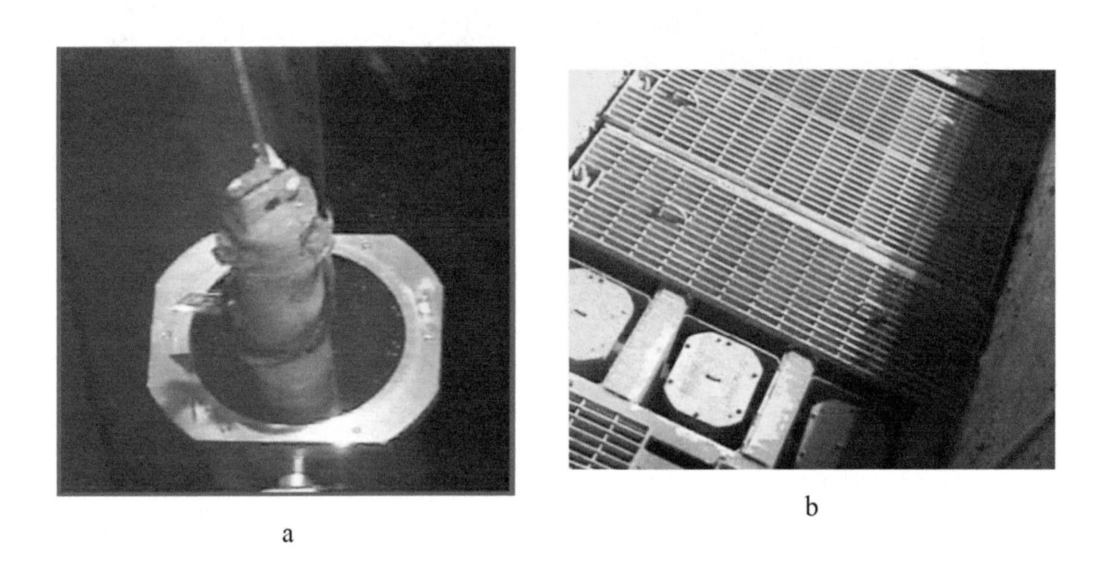

FIG. 16. (a) Fuel can being loaded into oversized storage and (b) oversized storage racks (courtesy of the Savannah River National Laboratory, USDOE [108, 113]).

7.1.2. Good practices at the L-Basin

The wet storage experience at SRS has been extensively documented and has helped shape international programmes on corrosion evaluation and control. This is reflected in a number of IAEA publications, such as Technical Reports Series No. 418, Corrosion of Research Reactor Aluminium Clad Spent Fuel in Water [72], IAEA-TECDOC-1637, Corrosion of Research Reactor Aluminium Clad Spent Fuel in Water [7], IAEA Nuclear Energy Series No. NP-T-5.2, Good Practices for Water Quality Management in Research Reactors and Spent Fuel Storage Facilities [18], and Ref. [14].

7.1.2.1. The L-bundles

As explained in Ref. [45], airborne dust in a spent fuel storage facility settles on reactor pools or spent fuel basin surfaces. The concept of the L-bundle is an improvement for wet storage of RRSNF because it minimizes settlement of solids on the surface of the fuel cladding, which in turn decreases the pitting corrosion process [7]. Thus, the integrity of cladding is maintained for longer periods, allowing also the extension of the wet storage of the RRSNF.

7.1.2.2. The machinery basin

The L-Basin has a machinery basin, an area with some underwater equipment required for the inspection, disassembly and bundling of fuel assemblies. The equipment includes a tilt table and an underwater saw located in working areas called canal stations that range in depth from 5 to 9 m, and three pits that are each 15.55 m deep. Facility personnel on walkways above the basin water remotely operate the equipment.

Empty fuel L-bundles are secured to the tilting table in horizontal position, and the SNF assemblies are placed into them, as shown in Fig. 17 (a). The table is slowly rotated to the vertical position after each assembly is inserted. After a designated number of assemblies have been placed into the bundle, the tilting table is lowered and a lid with a handling bail is secured to the bundle. The bundle is then transferred via a manual chain hoist to the VTS area.

If cropping of the fuel assembly is required to fit into the L-bundle or to maximize storage capacity, the assembly is first placed on the saw table shown in Fig. 17 (b) and the non-fuel end pieces are removed before insertion of the fuel element into the L-bundle. The end pieces are transferred to scrap buckets for later disposal.

7.1.2.3. The water chemistry control system

Ion exchange resins, sand filter trains and zeolite trains have been used to control the water chemistry at the L-Basin, as explained in Section 2.3.1. The sand filter removes insoluble particles, whereas the ion exchange resins remove and replace deleterious ions, minimizing corrosion. The zeolite train has been used to remove caesium and strontium ions [115].

Activity limits for the resin train are established to meet the safety basis for the facility. These limits translate into basin water operational activity limits for caesium and alpha activity as shown in Table 4, which also specifies the operational limits for pH, electrical conductivity and ionic species in the L-Basin, in addition to the monitoring frequency established for each parameter.

a

b

FIG. 17. (a) Closing an L-bundle in the machinery basin, which also has (b) a saw machine to crop non-fuelled parts of RRSNF (courtesy of the Savannah River National Laboratory, USDOE [114]).

TABLE 4. L-BASIN WATER OPERATIONAL LIMITS [116]

Water quality parameters	Operating limit	Monitoring frequency
pH	5.5–8.5	Weekly
Electrical conductivity	10 μS/cm	Weekly
Specific activity	^{137}Cs: 8.33 Bq/mL (500 dpm/mL) Alpha: 0.05 Bq/mL (3 dpm/mL) Tritium: 14 800 Bq/mL (0.4 μCi/mL)	Weekly Monthly Every 6 months
Concentration of Cu	0.1 ppm	Every 6 months
Concentration of Hg	0.014 ppm	Every 6 months
Concentration of Cl	0.1 ppm	Every 6 months
Concentration of Fe	1.0 ppm	Every 6 months
Concentration of Al	1.0 ppm	Every 6 months
Temperature	45°C	Monthly

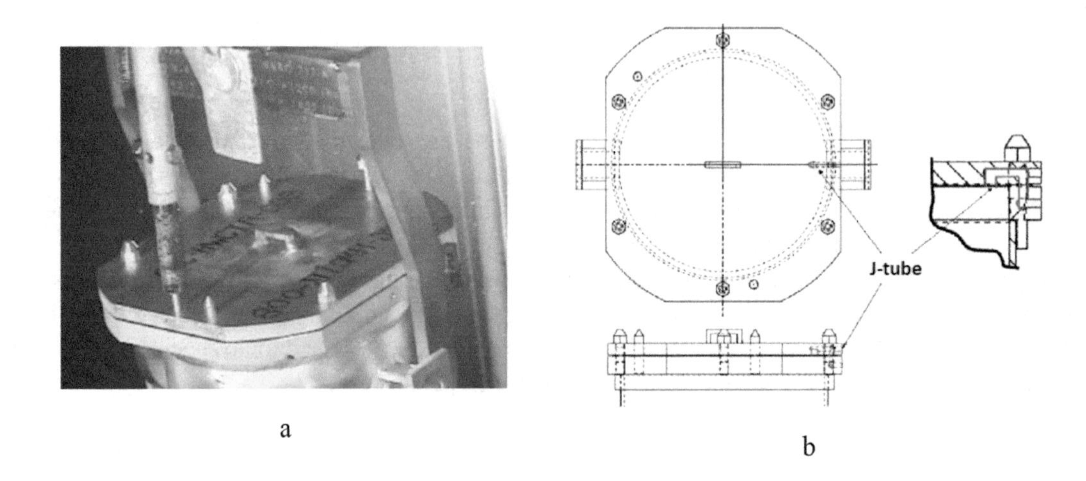

FIG. 18. Oversized storage can (a) lid and (b) J-tube (courtesy of the Savannah River National Laboratory, USDOE [113]).

7.1.2.4. The oversized storage cans

The oversized storage cans were designed to sequester the radionuclides released from the damaged fuel they contain. Inside the oversized storage can, the gases released from the damaged fuel (through its failed cladding) build up at the top of the can and into a designed 'J-tube' on its lid (Fig. 18). This evolved gas helps separate the internal water environment from the bulk basin water, ensuring not only the radiological protection of the personnel but also optimizing the water chemistry control, since it minimizes the need to replenish the resin beds.

The water within the oversized storage cans may become increasingly more contaminated with fission products. During the unloading activities at the RBOF for the transfer of the SNF into the L-Basin, it was necessary to open the oversized storage cans, which required the development of a system capable of locally removing the highly contaminated water and cleaning it within the oversized storage can, keeping it separate from bulk basin water [117]. A submersible underwater deionizer system (see Section 7.1.2.5) was developed for this purpose.

7.1.2.5. *The underwater deionizer system*

The underwater deionizer system consists of an ion exchange column, redundant pumps with their motors and a filter [117]. Its dimensions were minimized to enable as much flexibility and mobility within the basin as possible, 4.6 m below water surface. The ion exchange column removed radioactive material from oversized storage can water with the exact amount of cation resin that ensured maximum ion exchange efficiency. The water was pumped from the oversized storage can into the column by two redundant pumps, preceded by a filter and then each followed by a check valve to avoid recirculation (Fig. 19). During its operation, the system provided excellent results: the dose rates in the general basin water were maintained below the limit of 0.02 mGy/h [115].

7.1.2.6. *The corrosion surveillance programme*

Since 1992, SRS implemented a corrosion monitoring programme. At the time, the emplacement and removal of corrosion coupons demonstrated the poor water quality, but this motivated and later demonstrated the effectiveness of the water chemistry improvements made by the Disassembly Basin Upgrade Project in 1994–1996 [118]. The corrosion surveillance programme demonstrated that the aggressive corrosion attack in the basins of SRS had been eliminated by 2001 (Fig. 20) and continues providing evidence of the maintained good water quality [116, 119].

The extensive experience of SRS with the corrosion surveillance programme has demonstrated the importance and benefits of a continuous monitoring of water chemistry and corrosion, shaping international programmes as the ones mentioned in the introduction to Section 7.1.2.

7.1.3. Conclusions

At SRS, systems to prepare fuel, isolate damaged fuel, and to monitor and maintain water quality conditions of storage areas have been established and continuous evaluation of the fuel condition and characteristics has demonstrated their effectiveness. To ensure reliable safe storage throughout the storage period of SNF, continued improvement in water quality combined with a surveillance programme for evaluation of the fuel and storage system materials remains paramount. According to engineering judgement that is based on available literature, it is understood that the present management programme

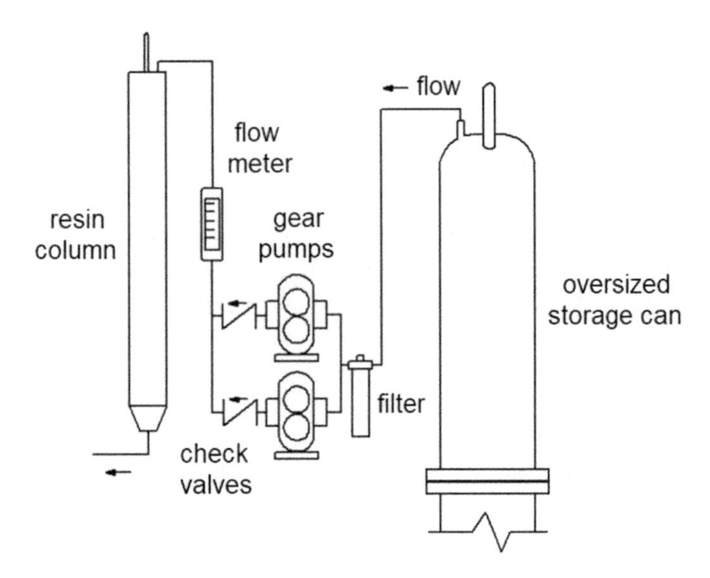

FIG. 19. Underwater deionizer flow diagram (courtesy of the Savannah River National Laboratory, USDOE [115]).

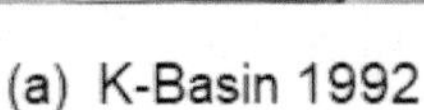

(b) K-Basin 2001

FIG. 20. Specimens used at the K-Basin as part of the corrosion surveillance programme in SRS, stored for (a) 12 months in poor water quality and (b) 45 months in improved water chemistry (courtesy of the Savannah River National Laboratory, USDOE [18]).

will provide full retrievability of the stored fuel for an additional period greater than 50 years, the projected lifetime of the basin structure.

The wet storage experience in SRS enables understanding of the effects of water chemistry on material performance and demonstrates that well-founded technologies and activities like the ones described above ensure safe wet storage of new, damaged and long term stored SNF. As a result, the L-Basin case continues to be referenced as an example of good practice by local and international programmes and guidelines.

7.2. CENTRALIZED DRY STORAGE AT HABOG

This case study demonstrates the recognition of the HABOG facility, operated by COVRA, as an example of good practice in dry storage of RRSNF. COVRA, founded in 1982, is responsible for managing all categories of radioactive waste in the Netherlands. In 1984, a government report concluded that in the Netherlands the waste could be stored in buildings for at least 100 years and a deep geological disposal facility would be constructed during that period, to be implemented after the storage [54]. At the COVRA site, four storage buildings contain the radioactive waste of the Netherlands, and in particular, the HABOG contains the HLW, including the RRSNF [120].

Construction of the HABOG facility started in 1999, and in September 2003, the facility was officially inaugurated [121]. A few months later, in November 2003, HABOG was ready to receive the first package of RRSNF. Figure 21 shows the location of HABOG within the COVRA complex.

7.2.1. HABOG facility description

The facility (90 m long, 45 m wide and 20 m high) was designed to last at least 100 years, built with highly reinforced limestone based concrete to withstand 15 different design basis accidents including earthquakes, an aircraft crash, flooding, outside explosions caused by a passing ship loaded with liquid gas, windstorms of 125 m/s, explosions in the facility, fire, hurricanes and a drop of a package from a crane during handling. The robustness of the construction of the building includes 1.70 m thick reinforced walls to ensure that none of these accidents would result in a significant radiological impact, preventing any consequences for the population and the environment [121].

HABOG is a vault type storage facility divided into three separate compartments. The first compartment is used for reception and unloading of the transport casks with HLW and spent fuel. The second compartment has three bunkers, specifically designed for the storage of canisters and other packages containing HLW that does not need to be cooled (e.g. CSD-C package, hulls, and cropped extensions of fuel extremities and other HLW). The third compartment also has three vaults which are used for the storage of vitrified HLW from reprocessed spent fuel originating from the Dutch NPPs, for RRSNF originating from the two research reactors, and spent uranium targets from molybdenum production [120].

As a standard procedure in the HABOG facility, the RRSNF is typically received in MTR2 transport casks, similar to the one shown in Fig. 22. Once the cask is removed from the truck it is transported to the treatment area where the basket is transferred from the transport cask to a stainless steel canister and welded tight in a hot cell. The canister is then filled with inert helium gas up to 154 kPa and checked for leaks. The leak criterion is 10^{-4} Pa • L • s^{-1}. After checking for leaks and contamination, the canister is placed in the storage vault [120].

FIG. 21. HABOG within the COVRA complex (courtesy of COVRA N.V.).

a

b

FIG. 22. Handling the transport cask in HABOG. (a) Unloading and (b) lid welding in the hot cell (courtesy of COVRA N.V.).

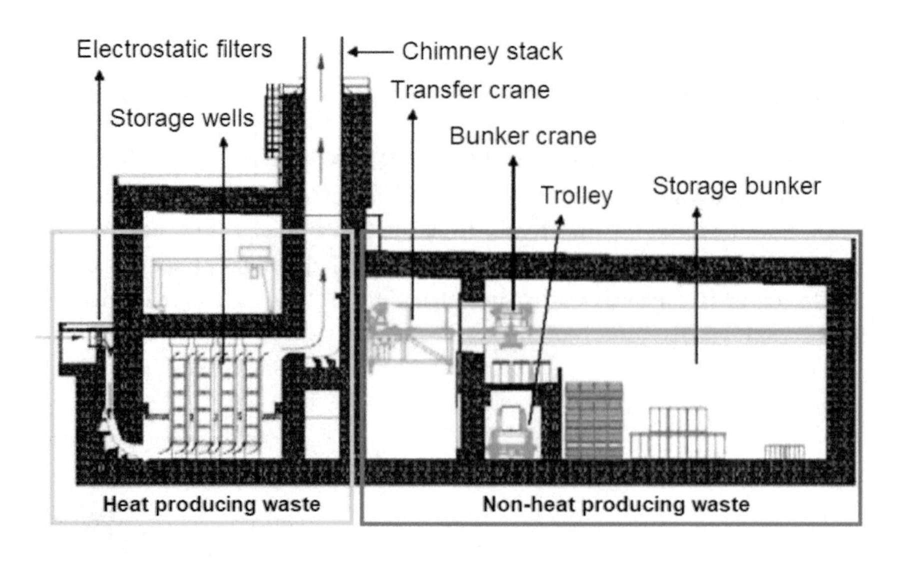

FIG. 23. A cross-section view of HABOG (courtesy of COVRA N.V. [121]).

In the storage vault, the RRSNF, spent uranium targets and vitrified HLW are stacked on five levels in vertical, air cooled storage wells. The storage wells are filled with an inert gas to prevent corrosion of the canisters and are equipped with a double jacket to allow passage of cooling air. The double jacket ensures that there is never direct contact between spent fuel, spent targets, or waste canisters and the cooling air. The cooling system is based on natural convection [121]. A schematic of the storage compartment for spent fuel and vitrified HLW is shown in Fig. 23.

The complete cycle for the treatment, transfer and conditioning of a MTR2 transport container with 33 RRSNF elements takes about five days. For preparation, testing of the equipment, and cleaning the installation, another five days are needed, which makes, in total, ten working days per campaign. For treatment of a TN28 cask, with 28 canisters of vitrified HLW from reprocessing, the campaign takes about 15 days, including receipt and return of the transport flask [121].

7.2.1.1. Storage capacity

In November 2003, HABOG received the first spent fuel from the Petten High Flux Reactor, Netherlands, followed in 2004 by vitrified HLW received from a previous reprocessing campaign in France, and spent fuel elements from the Hoger Onderwijs Reactor in Delft, Netherlands [120]. In total, one TN28 cask with 28 flasks containing vitrified HLW arising from the reprocessing of spent fuel from the Borssele NPP and eight MTR2 containers containing RRSNF were received in 2004 [121].

Since the start of operation until 2013, 26 MTR2 containers with RRSNF, four MTR transport casks with uranium filters from the molybdenum production factory, and seven TN28 flasks, each containing 28 canisters of vitrified HLW originating from the reprocessing of spent fuel from the NPPs in the Netherlands were unloaded, packed and stored in the HLW section of HABOG [48].

According to the licensee granted, COVRA must be able to unload a complete vault or bunker for inspection. This means that only two of the three bunkers and two of the three vaults may be completely loaded [121]. Thus, to meet this requirement, the design capacity of HABOG is limited to:

— 270 canisters with vitrified waste from reprocessing of spent fuel;
— 70 canisters each containing 33 spent fuel elements;
— 600 drums with HLW, non-heat generating.

In terms of volume, this is equivalent to about 750 m^3, which corresponds to less than 1% of the total volume of the HABOG building.

The storage capacity of HABOG was originally based on the operational lifetimes of the NPPs in the Netherlands, Borssele and Dodewaard, and the research reactors. Dodewaard was permanently shut down in 1997, and the closure of Borssele was originally planned for 2003. However, after construction of HABOG, the government decided to keep Borssele operational until end of 2033, which requires an expansion of HABOG to store the additional HLW resulting from the additional reprocessed fuel. As a result, two more storage modules for heat generating HLW need to be constructed. Currently, the extension of the HABOG is in progress. Design started in 2016, construction began in 2017 and completion is expected in 2021. The extension will offer 50 m^3 of extra storage capacity for HLW. At present, approximately 110 m^3 of HLW is stored in the HABOG [120].

7.2.2. Good practices at HABOG

The infrastructure within HABOG allows the handling of HLW, in air, without jeopardizing the safety of operational workers. The hot cell available in the facility, shown in Fig. 24, has the entire infrastructure needed to remotely unload the basket from the MTR2 cask, transfer it to the stainless steel storage canister, and seal the top lid without exposing workers to radiation. Precautions taken during construction limit the number of places that may be contaminated with radioactive material due to the handling of RRSNF or HLW, and the finishing of all surfaces in places where radioactive material is handled was performed in such a way that any radioactive contamination can be easily removed. To date, no contamination has been detected in the hot cell.

The selected transport casks and use of removable gates and walls of the cells provide an efficient radiological shielding. In 2005, a dose rate of 5 μSv/h was measured during a campaign involving a TN28 cask. This dose rate was about a third of the dose rate foreseen during the design phase of the facility; based on calculations, the latter dose rate was expected to reach around 17 μSv/h [121].

A safety feature in HABOG is its mechanical ventilation system, which is designed to keep the building at a slight positive pressure (except for the vaults for spent fuel and HLW). By doing this, the air flow through the building is directed from areas with no contamination towards areas with a potentially higher contamination. Both incoming and outgoing air is monitored and filtered.

FIG. 24. Process hot cell HABOG (courtesy of COVRA N.V.).

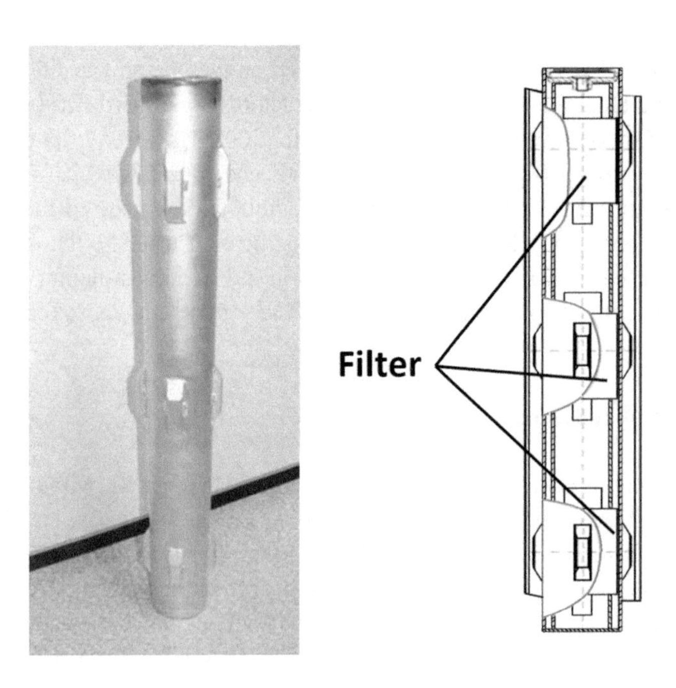

FIG. 25. Aluminium filter bin for three uranium filters (courtesy of COVRA N.V. [122]).

7.2.2.1. Waste from multiple streams

A unique feature of the HABOG facility is its ability to comingle containers of SNF with those of uranium filters from the molybdenum production factory. The uranium filter contains the remains of uranium targets used for the production of ^{99}Mo. The filters are placed in an aluminium filter bin (three filters per container) and closed by welding (Fig. 25) [122]. The dimensions of the aluminium filter bin are comparable to those of SNF elements. With this pragmatic approach, the same transport equipment and tools can be used.

7.2.2.2. The passive cooling system

One of the most conspicuous features in the design of the HABOG facility is the application of the natural convection concept for the control of the temperature of the spent fuel and HLW canisters, as shown in Fig. 26. Natural convection was chosen because of its inherent safety characteristics: to ensure cooling under conditions of loss of electric power and to be insensitive to human errors [120].

In the storage vault, canisters of SNF and vitrified HLW are stacked vertically on five levels in vertical air cooled storage wells. The storage wells are filled internally with an inert gas to prevent corrosion of the canisters and are equipped with a double jacket to allow the passage of cooling air. The double jacket ensures that there is no direct contact between spent fuel or radioactive waste canisters and the cooling air. The vault system of HABOG is shown in Fig. 27. Containment and storage wells are continuously monitored for temperature and radiation levels.

7.2.2.3. The COVRA public communication programme

Another remarkable aspect of HABOG and COVRA is the public communication programme developed to attract the public to visit the facility. Since the beginning, attention was paid to psychological and emotional factors in the design of the technical facilities. It was felt that a good-looking exterior, along with a good working atmosphere and being open to visitors could help to establish a good relationship with the public.

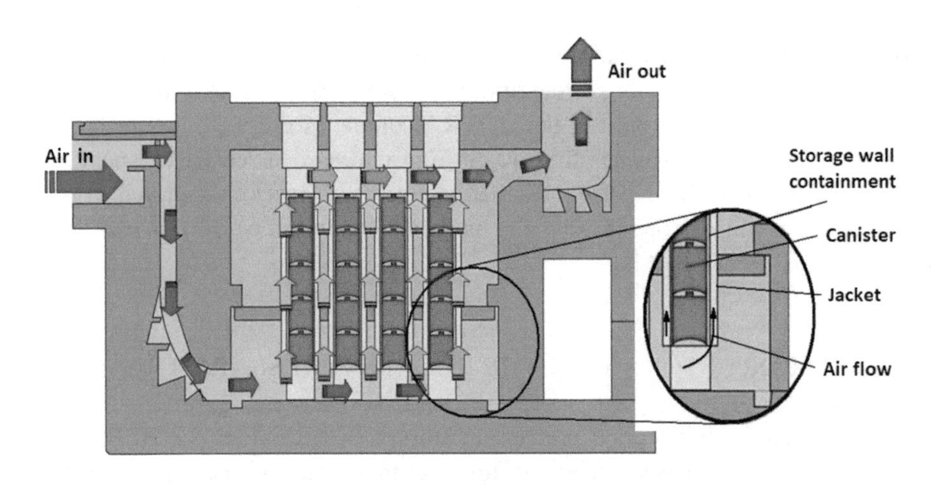

FIG. 26. The passive natural circulation cooling concept used in HABOG (courtesy of COVRA N.V.).

FIG. 27. The vault system of HABOG showing (a) the storage room during construction, (b) the current top view, and (c) a bottom view of the storage tubes (courtesy of COVRA N.V.).

The whole HABOG building has been integrated into an artistic concept comprising representative artwork inside and a carefully chosen façade outside. The exterior of the building is to be repainted every 20 years in a slightly lighter shade, reflecting the decay of the heat production of the waste inside, so that after 100 years, the colour of the outside walls is white. During office hours, the main gate is open, showing openness, even though of course, the whole site is permanently guarded. Visitors are welcome to the facilities, and guided tours are available which allow the public to look at the work being done. More detailed information about the communicative approach of the HABOG facility can be found in Ref. [121].

7.2.3. Conclusions

In the Netherlands, the government's policy mandates that all radioactive waste must be stored above ground in engineered structures allowing retrieval at all times, for a period of at least 100 years.

The policy is based on a step-wise decision process in which all decisions are made to ensure safe disposal without excluding new solutions in the future. To comply with this requirement, the multipurpose HABOG facility was designed and built in the COVRA complex for dry storage of HLW and RRSNF, welded tight in canisters. The heat generating waste from the reprocessed HLW, spent uranium targets used for ^{99}Mo production and RRSNF from the national research reactors is stored in naturally cooled vaults. The facility was commissioned in 2003 and has been recognized worldwide as a good example for interim dry storage of HLW and RRSNF.

7.3. WET AND SEMIDRY STORAGE AT THE BUDAPEST RESEARCH REACTOR

This case study demonstrates good practice in storage of RRSNF at the BRR. The organization operating the BRR developed a new technique for interim storage of spent fuel, keeping it dry and using the pool water as shielding and a heat sink to remove the decay heat produced. The details of the technology development and its implementation can be found in Ref. [123].

7.3.1. Description of the Budapest Research Reactor

At the Central Research Institute of Physics (known locally as the KFKI) campus in Budapest, Hungary, the BRR has been operated by the Atomic Energy Research Institute (known locally as the AEKI) since 1959. The BRR is a tank type research reactor, moderated and cooled by light water. In 1967, it underwent a first power increase (from 2 to 5 MW) along with a change in fuel (from EK-1 to VVR-SM) and beryllium reflector. In 1986, it began a full-scale reconstruction, which included a power increase to 10 MW, another change of fuel (to VVR-M2), and it commenced its licensed operation in 1993. The BRR has two storage pools for SNF, the AR pool and the AFR pool, located inside and outside of the reactor hall, respectively.

The 600 m^2 reactor hall is sealed hermetically and ventilated individually, with the concrete reactor block at the centre. Located near the reactor block is the AR pool (Fig. 28 (a)): a steel tank filled with deionized water, with storage racks that can accommodate 786 VVR-shaped assemblies. The AR pool's decay heat is removed by a cooling pipeline and its water is filtered by a mobile ion-exchanger when the electrical conductivity reaches 3 μS/cm. The AR pool is served by the reactor hall crane and through the two service plugs located in the iron plates that cover the pool. The service plugs can accommodate the container used to transport the fuel assemblies among the reactor, the AR pool and the AFR pools. The AR pool holds unloaded SNF, the beryllium reflector elements and any damaged fuels, and ensures place for a complete core unloading in case of emergency. Thus, from the 786 storage positions, 230 have to remain empty all the time for the core unloading, 216 are used for the beryllium reflectors, and 5 are allocated for any failed or leaking fuel assemblies (sealed within in a special wet container [124]), which limits the effective capacity of SNF storage.

The original SNF management strategy involved repatriation of the SNF to its country of origin, the former Soviet Union. However, the negotiations on SNF shipments took longer than expected, and considering the limited capacity of the AR pool, it was decided to construct the AFR pool during the first upgrade of the reactor.

The AFR pool is a cylindrical stainless steel tank (Fig. 28 (b)) embedded below ground level in a monolithic reinforced concrete block, about 100 m away from the reactor hall. The AFR pool has racks that accommodate 188 aluminium tubes, which in turn serve as storage locations for the fuel elements. The AFR pool is also filled with deionized water, which is filtered by a mobile ion-exchanger when the electrical conductivity reaches 10 μS/cm. The AFR pool is served through a service plug located above the reception seat. It was constructed in 1967 with a metal-framed building over it and a removable roof. However, in 2007 the AFR pool was remodelled and a service hall was erected around it, and the original 8 tonne angle-type crane was changed to a 15 tonne bridge crane and a transfer trolley.

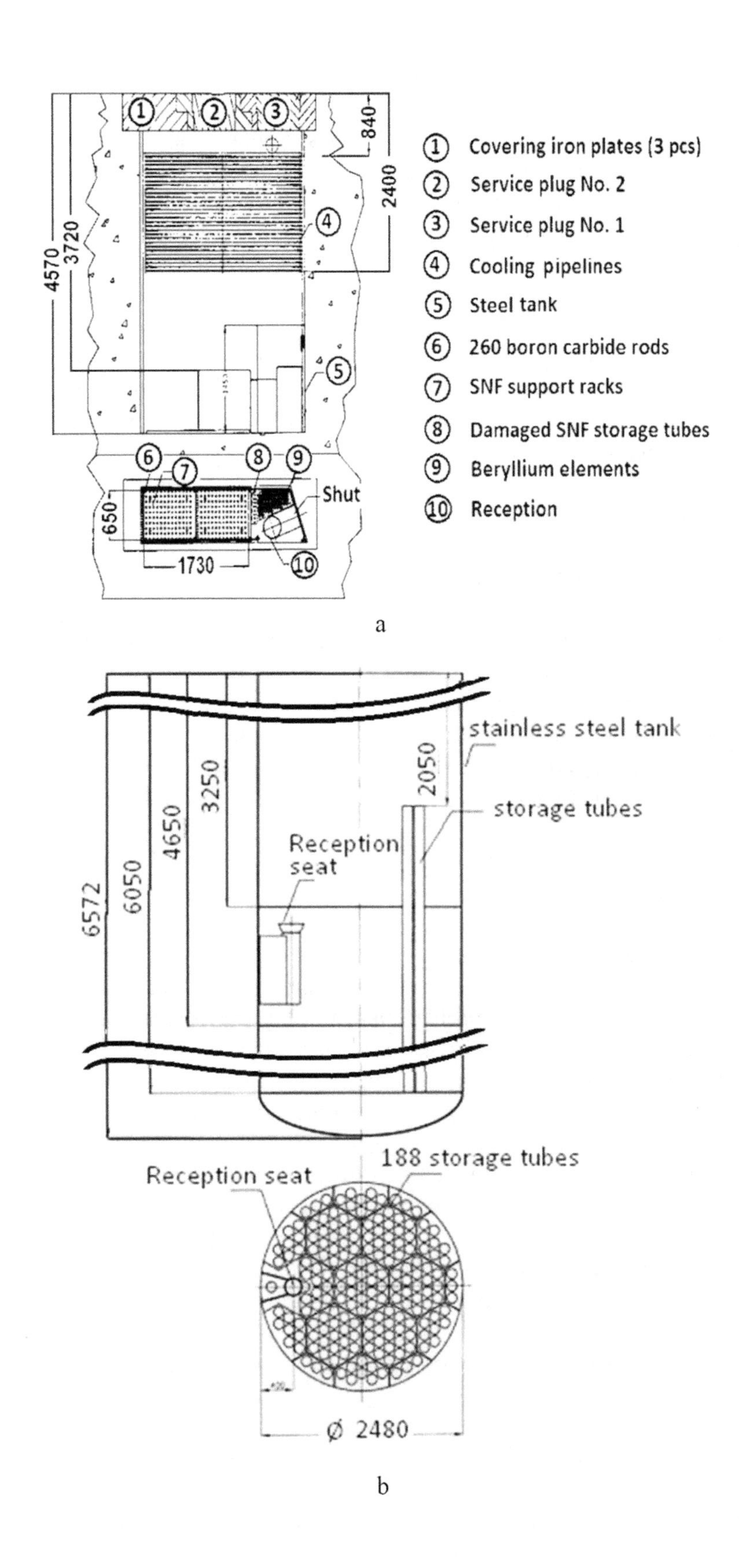

FIG. 28. The BRR (a) AR pool and (b) AFR pool (courtesy of BRR).

7.3.2. Good practices at the Budapest Research Reactor

The BRR includes the water quality control and surveillance procedures of both pools in its operation manual. Most of the parameters (electrical conductivity, pH, ^{137}Cs, ^{60}Co, corrosion products) are controlled by periodic sampling technique, and only water level, water temperature and exhausted air activity are measured continuously (online). The summary of the water quality parameters and the frequency of the inspections are given in detail in Ref. [123].

Even though there was a lack of experience of long term wet storage, it was well known that corrosion of aluminium can advance rapidly once it has started, and by 1995 some of the stored assemblies were almost 40 years old. Consequently, a visual inspection programme was implemented, with the objective of comparing the condition of cladding topography each decade. The first visual inspection was made in 1999, using the Hungarian Underwater Telescope, and the results showed corrosion in some fuel assemblies, at different levels.

7.3.2.1. The semidry storage technology

The long period in which RRSNF was kept in wet storage was the main reason that BRR developed the semidry technology. Encapsulation of the stored SNF assemblies in an inert gas atmosphere was meant to slow down or even stop the corrosion process. The capsules were stored in the pool, using water as biological shielding and a heat sink.

In addition to slowing down the corrosion process, extending the interim storage period of the spent fuel through the encapsulation technology allowed an increase in the capacity of the AFR storage pool by one third of its original capacity (1692 to 2256 fuel elements). Without canning, the spent fuel had to be stored in three levels on the aluminium tubes. The canning technology allowed the storage of the spent fuel in four levels.

7.3.2.2. The encapsulation process

Tube-type capsules, 3 mm thick, made of aluminium alloy were used to encapsulate the SNF. The capsule can accommodate one EK-10, one triple VVR type, or three single VVR type assemblies, as shown in Fig. 29 (a). The encapsulation machine shown in Fig. 29 (b) was used to submit the capsule (with the SNF placed inside) to a powerful drying procedure, heated by an eddy current coil; the capsule was then filled with dry nitrogen gas, and finally closed by the welding on of a capsule head. The machine was mounted in the service plug of the AFR pool, so that the capsule was submerged about 50 cm under water, ensuring safe manipulation.

Almost all the steps of the encapsulation process were performed automatically by the machine in the vacuum sealed operation chamber, which contains five service blocks:

(1) Transfer of the capsule into the operation chamber;
(2) Removal of the water from the capsule;
(3) Drying of the SNF;
(4) Vacuuming the capsule, filling it with nitrogen and pressing the capsule head into the capsule;
(5) Welding the capsule head.

The detailed description of the encapsulation process, the machine and the capsules can be found in Ref. [125].

Figure 30 shows the upper part of the capsule, with the capsule head securely welded to the capsule. The encapsulation process took around 120 minutes, which permitted processing of about five spent fuel assemblies per working day. The process was licensed in 2002, and encapsulation of all SNF discharged prior to 1986 was completed in 2004.

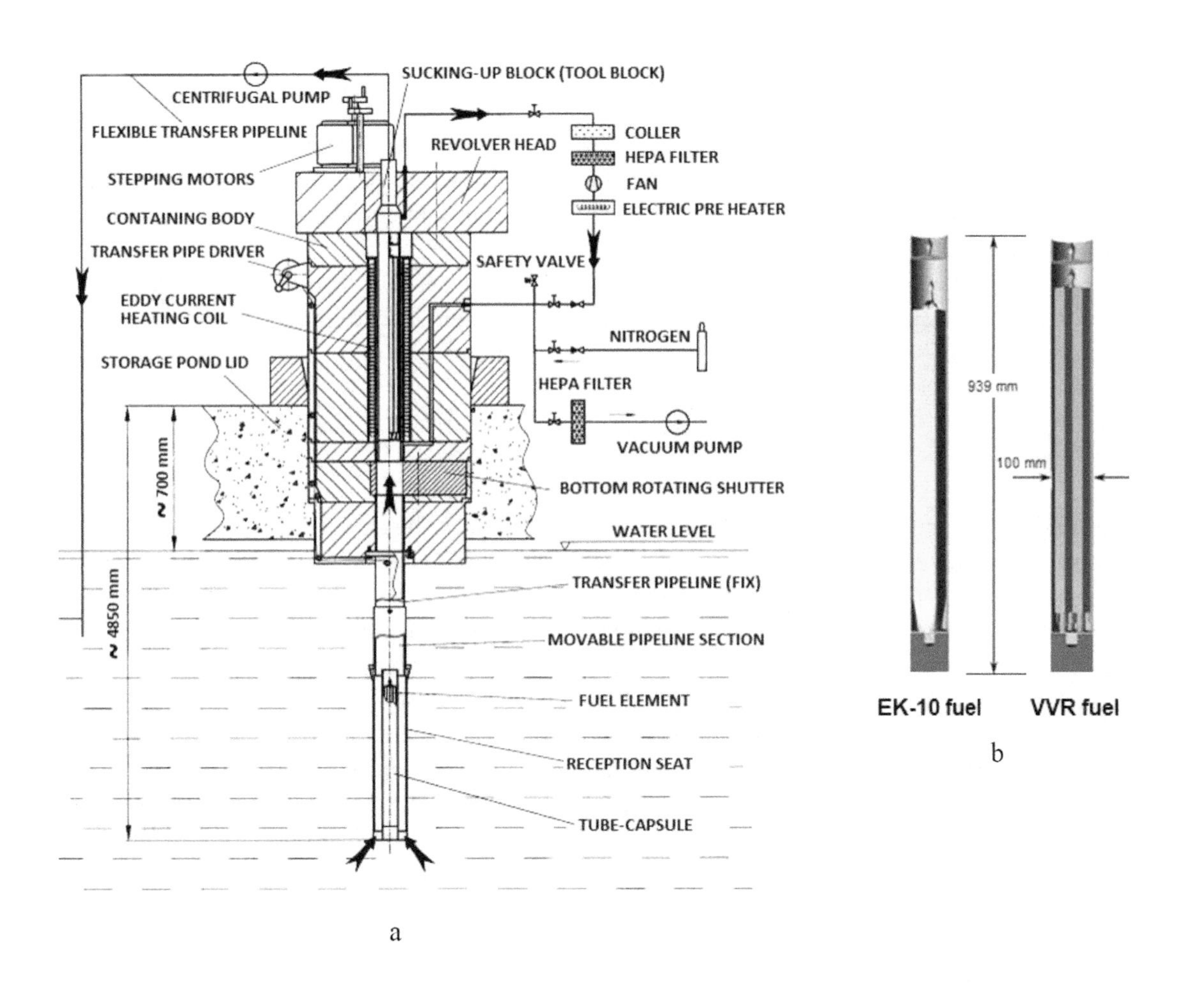

FIG. 29. (a) Conceptual layout of the encapsulating machine and (b) the encapsulation tube (courtesy of BRR [123]).

FIG. 30. Canning tube secured by welding (courtesy of BRR [123]).

7.3.2.3. The return of the spent fuel to the Russian Federation

With the implementation of the Russian Research Reactor Fuel Return programme, all SNF assemblies irradiated in the BRR before 2005 were eligible to be returned to the Russian Federation. Therefore, all the assemblies in the AFR pool, both in wet and in semidry storage, were shipped to the Russian Federation in 2008.

A part of the shipment agreement was that the 342 semidry capsules had to be opened, each fuel assembly was visually examined, then only the fuel assemblies were loaded into the transfer containers. The examination of the semidry assemblies, along with the examination of the empty AFR pool, resulted in the conclusions found in the next section.

7.3.3. Conclusions

No fission products were found in the AFR pool water, nor in the sludge. This was a result of the proper storage procedures in the pool. The 40 year wet storage period (1968–2008) was possible thanks to the good water quality and its proper maintenance.

The encapsulating machine and its accessories represented a compact and mobile system that ensured an almost completely automatic, safe and reliable encapsulation of SNF. This novel technique provides an alternative for the research reactor community facing SNF storage issues or wishing to extend the safe storage of their SNF until suitable disposal options are put in place.

Only five years passed between the encapsulation of the SNF and the opening of the capsules, but the visual inspection proved that the semidry storage of the fuel in the inert gas atmosphere stopped cladding corrosion, which could have allowed at least 50 more years of safe storage, even in the case of fuels that had exhibited significant corrosion prior to their encapsulation.

7.4. WET STORAGE REMEDIATION AT VINČA INSTITUTE OF NUCLEAR SCIENCES

This case study demonstrates the consequences of prolonged spent fuel storage in water of improper quality. It is not the purpose of this publication to provide a root cause analysis of the improper water quality, but to provide discussion of the technical aspects, focusing on the problems that arose from the total abandonment of the water treatment, and the possible implications for safety.

7.4.1. Description of the RA research reactor

The RA research reactor at the Vinča Institute of Nuclear Sciences, Yugoslavia (now Serbia), is a Russian type, 6.5 MW heavy water reactor. The reactor was designed to operate with LEU metal fuel. It reached first criticality in December 1959 and started full power operation by the end of 1960, after a short period of initial testing. In 1976, the reactor fuel changed from LEU fuel elements (2% of ^{235}U) to HEU fuel elements (80% of ^{235}U) with the same slug geometry, but in the form of UO_2 dispersed in an aluminium matrix.

The RA facility was built with a 200 m^3 storage pool, consisting of four interconnected rectangular basins and an annex to the fourth basin. Underwater transfer of the SNF assemblies from the reactor pool to the storage pool was possible since the two pools are connected by a transport channel.

Figure 31 shows a diagram of the storage pool and its basins which are 6.5 m deep and have a volume of around 40 m^3 each. The concrete walls and bottom of the basins are lined with 1 cm thick stainless steel plates. The water level in the basins is around 70 cm below the top of the basins (ground level) and ventilation ducts were installed in this part of the basins' walls, above water.

The openings of the basins at ground level are covered by movable carbon steel plates. The SNF assemblies were held with stainless steel channel holders (SSCHs) which, in turn, were positioned in the basins by a carbon steel grid. A grid for each basin was placed above the water level, under the basins'

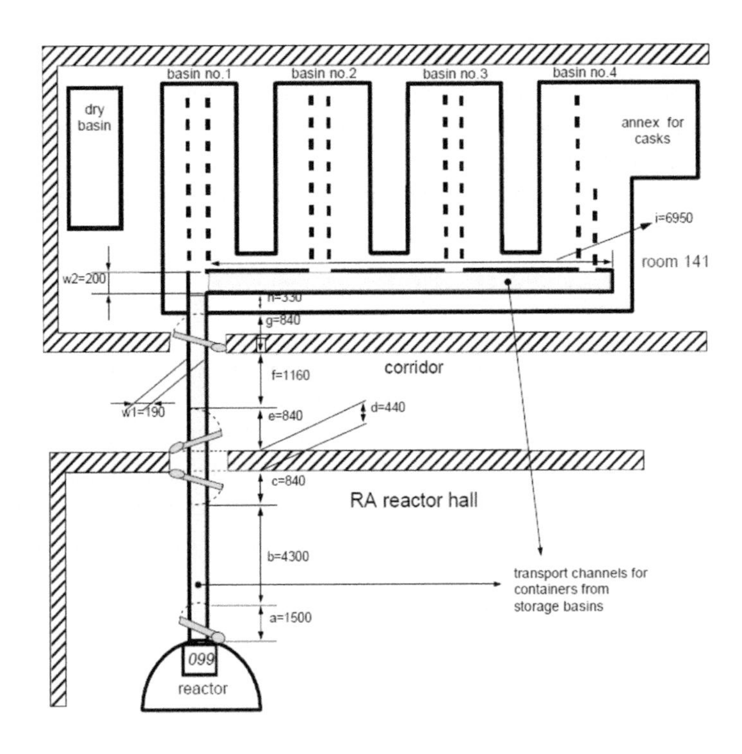

FIG. 31. Diagram of the RA reactor spent fuel storage pool [7].

carbon steel covers. Exact dimensions and detailed technical descriptions of the storage pool and basins can be found in Ref. [7].

The storage capacity of this storage pool is not well defined, given that it accommodates fuel assemblies with different enrichments. However, in the mid-1960s it was decided to increase storage capacity with the design and production of aluminium barrels, which would store 30 aluminium tubes, each with up to 180 spent fuel elements (fuel slugs). To ensure that the aluminium barrels remained subcritical, cadmium strips were inserted into them.

In 1984, the reactor was temporarily shut down for modernization and partial refurbishment of control and safety systems. However, due to various technical, legal, political and economic reasons, planned refurbishments had never been finished and the RA reactor was never put into operation again [126, 127]. In 2001, a proposal for permanent shutdown and decommissioning of the RA reactor was submitted to the government. Finally, in 2002, the government issued a directive to permanently shut down the reactor and approved the Vinča Institute nuclear decommissioning programme. Under this programme, three interrelated projects were established, supported through IAEA's technical cooperation programme [128–130]:

- — Safe removal of spent fuel arising from the RA reactor, and return of all fresh HEU fuel and spent HEU and LEU fuel to the Russian Federation;
- — Safe management of waste in the Vinča Institute;
- — Decommissioning of the RA research reactor through immediate dismantling.

At this point, the reactor had about 8030 spent fuel elements (corresponding to more than 2.5 tonnes of heavy metal), most of them contained in the four basins of the storage pool [128, 131]:

- — 6656 LEU spent fuel elements, of which 4929 were placed in aluminium barrels, and 1727 in SSCHs;
- — 1374 HEU spent fuel elements, of which 894 were placed in SSCHs, and 480 had been left in the drained reactor core.

In addition, the reactor had 5046 unused fresh HEU fuel elements. After being loaded into 17 TK-S15 and 10 TK-S16 casks, these fuel elements were repatriated to the Russian Federation in August 2002 under the Russian Research Reactor Fuel Return programme [132, 133]. Information on spent fuel management is provided in Section 7.4.2.3.

In 2019, the preparation of a strategic plan for the decommissioning of the RA reactor was still underway [134].

7.4.1.1. Initial condition of the storage pool

Unfortunately, water quality maintenance was not a concern in the AR pool of the RA research reactor during its operation and in the first years after its 1984 shutdown. According to the original design, the RA reactor spent fuel storage had no system for pool water purification [128]. Monitoring and maintaining of pool water radiochemical parameters were not imposed, since the pool water was not supposed to be in direct contact with the spent fuel. From the initial operation, the AR pool was filled with stagnant tap water, kept at ambient temperature, and compensation for any loss of water due to evaporation was done once a year by adding new tap water.

The first reference to poor quality water in the basin, based on visual inspection, is found in the 1962 Annual Report of RA Operation [135]. Water purification was recommended but no action was taken. The only change was the aluminium barrels that, after being filled with the spent fuel assemblies, were filled with demineralized water and then closed.

In the 1979–1984 period, the annual reports of the reactor [135] state that the visual examinations that were routinely made during core unloading procedures reported stains and surface discoloration on many spent fuel elements. At that time, these anomalies were attributed to inappropriate chemical parameters and reduced flow rate of the primary coolant and moderator in the reactor core (heavy water).

In 1984, low ^{137}Cs activity was detected in the storage pool water [136]. This time, however, the activity was attributed to leaks from a spent TVR-S LEU fuel element that had been laying at the bottom of the pool for around eight years. This 'lost' fuel slug was then removed, and it was proposed to install a water purification system and implement periodic monitoring of the chemical and radiation parameters of the storage pool water [137].

However, it wasn't until 1994 that actions started to be taken to improve the water quality of the storage pool. At this time, visual inspection of the pool revealed thick deposits of sludge on its walls and bottom, and it was determined that a thorough inspection of the storage facilities was needed, which resulted in a request for assistance from the IAEA, the USA, and the Russian Federation [7].

7.4.1.2. Inspection of the storage pool water

An IAEA fact-finding mission visited the Vinča Institute in September 1995 with main concern focused on the spent fuel storage pool [126]. The first analysis was performed on samples from the basins' water, the SSCHs, and the sludge at the bottom of both the storage pool and transport channel [7]. The water chemistry was inadequate for spent fuel storage: it was corrosive to aluminium alloys, the electrical conductivity and chloride content were high, and ^{137}Cs and ^{60}Co were detected. The stains and discoloration of spent fuels reported prior to the shutdown were now, in 1995, identified as aluminium hydroxide deposits that covered the aluminium cladding of the fuel slugs. The quantitative results of the measurements and analyses made at that time can be found in Ref. [7] but, as an example, the specific activity of ^{137}Cs was around 100 Bq/cm^3 in the pool water and around 1500 Bq/cm^3 in the sludge.

In 1996 and 1997, the Vinča Institute and the IAEA Seibersdorf Laboratories analysed the sludge samples in composition and activity and the water from the SSCH that had 'lost' the fuel slug [138]. It was found that the sludge was composed mostly of iron, and its main source was the corrosion of the carbon steel components (SSCH grid and basin covers) which produced iron oxide. As for the water of the SSCH that 'lost' the fuel slug, its ^{137}Cs specific activity was around 50 MBq/cm^3.

An example of the heavy corrosion suffered by carbon steel structures in this water can be seen in Fig. 32, which shows a structure (designed for decontamination of fuel assemblies) that stayed in basin No. 4 until its removal in 2006, and the appearance of the storage pool water due to such corrosion and dust.

7.4.2. Wet storage remediation activities

In 2002, Serbia joined a CRP on Corrosion of Research Reactor Aluminium Clad Spent Fuel in Water, initiated by the IAEA to study corrosion of aluminium alloys in RRSNF storage pools. The overall objective of the CRP was to improve management, storage practices and storage procedures at facilities used for interim wet storage of RRSNF, through better understanding of the localized corrosion of aluminium cladding and the ranges of water chemistry parameters that provide resistance to corrosion [7]. An important activity performed within the frame of the CRP was the exposure of circular samples (coupons of different aluminium alloys) to the water of the RA storage pool for periods of six months to six years [139, 140]. Figure 33 shows the conditions of some of the coupons after being exposed for several months to the storage pool water of the RA research reactor, and a bucket that was used to remove some of the sediment collected in the pool.

7.4.2.1. Partial cleaning of the at-reactor storage pool

Major activities started in 1997 to improve the condition of the storage pool and the quality of its water [7]. In 1998, most of the sludge and debris were removed and a special technique of sludge immobilization was developed at the Vinča Institute: metal barrels 200 L large were shielded with a thick concrete wall and then filled with about 75 L of a sludge–cement mixture (a sedimentation technique

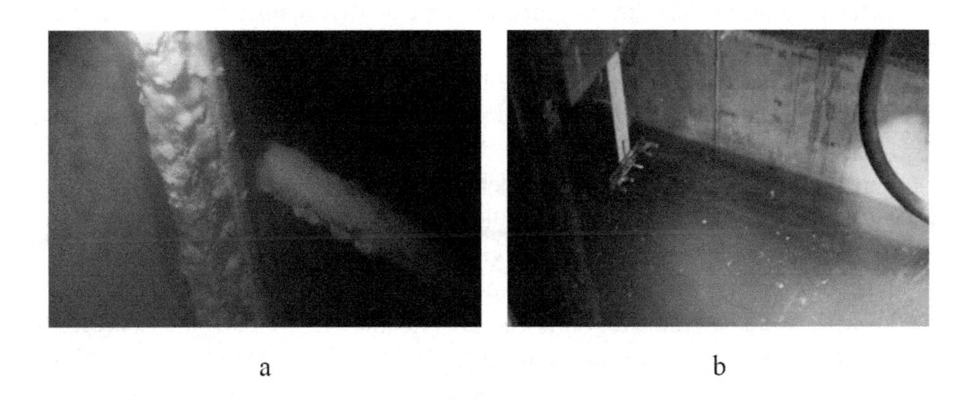

FIG. 32. (a) Corroded carbon steel structure inside of spent fuel storage pool and (b) floating dust on the surface of the AR pool [7].

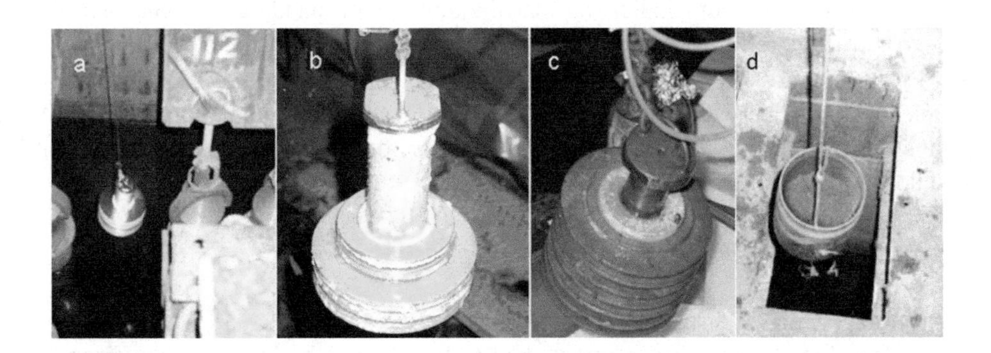

FIG. 33. (a) CRP test rack being immersed in the storage pool water then (b) removed after 24 months, and (c) removed after 72 months, as well as a (d) sediment collector [7].

also developed by the Vinča Institute). Around 40 of these barrels, having an average activity of about 150 MBq each, were stored as low level waste at the Vinča Institute [141, 142].

In 2004, it was reported (Ref. [139]) that even though pH and electrical conductivity levels remained only slightly improved, the visual clearness of the pool water had been regained thanks to the sludge removal and daily operation of a water purification system. The purification system consisted of around 60 L/min of water being pumped daily during working hours through a mechanical filter which retained particles bigger than 25 μm.

7.4.2.2. Inspection of the aluminium barrels and stainless steel channel holders

Keeping the fuel stored within aluminium barrels for more than 30 years raised concerns about an eventual increase in internal pressure from gaseous fission products released from leaking fuel and from hydrogen produced by radiolysis. The design pressure for barrel failure was determined as 1 MPa, enough to release about 20 L of gaseous matter (mainly hydrogen) into the pool [143]. In order to reduce the potential danger of spent fuel elements storage in barrels and to get information about the state of spent fuel elements, a joint Russian–Yugoslavian project was initiated in 1998 to perform underwater drilling of aluminium barrels. This project checked fuel cladding integrity, activity of water inside the barrels, and determination of probable high pressure inside the barrels, due to corrosion or fission gases. By the end of 2001, none of the 16 aluminium barrels drilled revealed high gas pressure. However, gas bubbles were observed when the aluminium barrels were shifted underwater. Very high specific activity of ^{137}Cs (0.5 to 1.5 MBq/cm^3) was measured in water samples taken from the drilled aluminium barrels. The lack of overpressure inside the barrels and very small gas activity indicated that the barrels were not leak-tight [143].

This led to the conclusion that not only had the first barrier against fission products release (the cladding) breached, but so had the second barrier (the aluminium barrel walls). It was also suspected that the cadmium strips, placed in the aluminium barrels to ensure subcriticality, could have reacted with aluminium from the barrel walls and fuel elements, contributing to increased corrosion inside the aluminium barrels. Fission product leaks (mainly ^{137}Cs nuclide) from failed fuel were confirmed [7]. All steps of the inspection process were done underwater, and after venting the barrel and getting the necessary water samples, the aperture in the barrel lid was sealed with a rubber plug, and the barrel was returned to its position.

In 2003, the ^{137}Cs and ^{60}Co specific activities, pH and electrical conductivity of water samples taken from about 200 SSCHs were determined [144]. The pH remained in the 7.31–8.15 range. The electrical conductivity was 50–400 μS/cm on average and up to 21 000 μS/cm for the most acid samples. The ion content was up to 70 mg/L for chlorides, around 30 mg/L for sulphates, and iron and aluminium ions were below detection limits of the analytical techniques used. The concentration of specific ions in water samples taken from the SSCHs was determined in only a few cases. This was due to the very high radioactivity level of the samples. About 10% of all water samples presented a very high ^{137}Cs specific activity, in the order of MBq/cm^3, indicating that the aluminium cladding of some fuel elements had breached due to corrosion, resulting in a probable leak of fission products from specific fuel elements into the SSCH [144].

Subsequent actions that were implemented for the storage pool remediation included the removal of the carbon steel structure in basin No. 4 (in 2006), the removal of the remaining sludge (in 2006 and 2007), continuous water purification and monitoring of its chemical parameters.

7.4.2.3. Spent fuel repackaging

Taking into account the structure of fuel elements and the containers in which they had been placed, it was clear that all spent fuel elements had to be repackaged into new containers suitable for transport. However, pulling out fuel elements from the aluminium tubes inside of the aluminium barrels would have been an impossible or very risky action that could damage the cladding, already weakened

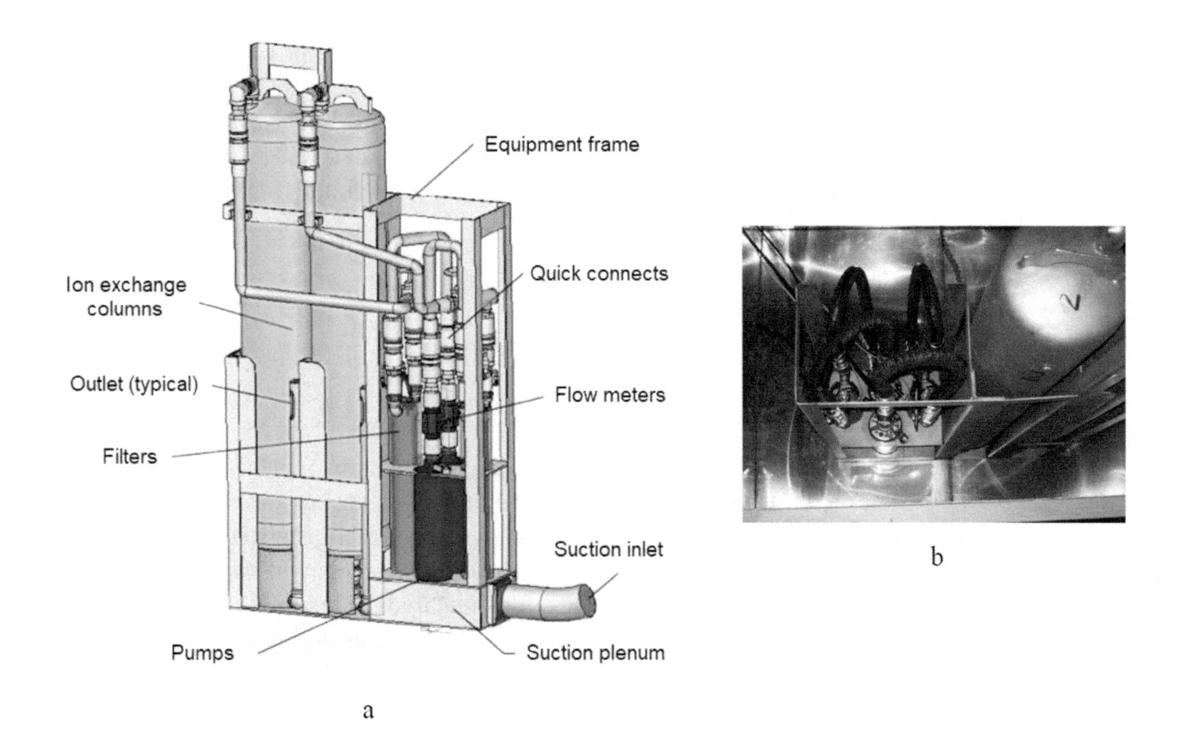

FIG. 34. (a) Diagram and (b) top view of the underwater deionizer of the water control system installed in the spent fuel storage pool of RA research reactor (courtesy of the Savannah River National Laboratory, USDOE [145]).

by the corrosion. Calculations revealed that upon opening of the aluminium barrels, some 10^{13} Bq of ^{137}Cs activity could be released into the spent fuel pool water, and that the same amount may be released within one year afterwards. Therefore, an efficient system for absorbing ^{137}Cs in the storage basins was mandatory if any fuel handling activities were to be performed there [131].

Because of the degraded condition of the storage pool and the building, a lot of facility modifications and new equipment provisions had to be implemented to enable repackaging and loading activities in the reactor building. These included: (i) removal of underwater metal structures from the spent fuel storage pond; (ii) replacement of the bridge crane in the reactor room and in spent fuel storage; (iii) upgrade and refurbishment of ventilation system, electric power supply system and control room in spent fuel storage; and (iv) acquisition of a 16-tonne capacity forklift, some radiation monitoring equipment and a water chemistry control system for ^{137}Cs removal, shown in Fig. 34 [145].

Finally, the entire inventory of spent fuel (see Section 7.4.1) was repatriated to the Russian Federation in November 2010, after being loaded into 16 TUK-19 casks and 16 SKODA casks [131].

7.4.3. Conclusions

It can be concluded that the prolonged use of water of improper quality, excessive amounts of floating debris on the pool surface, improper combination of metals, and inadequate control of both water composition and settled solids led to excessive corrosion of fuel cladding. This also led to cladding breach and contamination of the storage pool and the stored fuels, raising serious concerns about the safety of the installation.

As a consequence, technical activities that are required prior to spent fuel repatriation and that typically can be performed in less than six months, took more than three years in the case of the RA research reactor, at a cost higher than it otherwise would be if the water in the storage pool had been kept in good condition. Additionally, it resulted in around 40 metal barrels of low level waste that could have been avoided by keeping good quality water in the storage pool.

The lessons learned by the organization operating the RA research reactor can be summarized as follows: (1) use a good purification system to maintain good quality water in spent fuel storage pools; (2) avoid the use of plain carbon steel as a construction material for storage pools; (3) avoid the use of incompatible metals and alloys in the spent fuel storage pool; (4) prevent dust and debris from falling on pool surfaces; (5) avoid any accumulation of sludge in the pool; and (6) monitor corrosion.

8. CONCLUSIONS

Storage and eventual disposal of RRSNF poses major challenges to organizations operating a research reactor. Many States with one or more research reactor and no nuclear power programme will face the problem of dealing with the relatively small amounts of spent fuel generated by their research reactors. It is expected that the majority of States that operate research reactors will, for different reasons, defer selecting an ultimate solution for their RRSNF. Therefore, hundreds of research reactors worldwide, both operational and shut down but not yet decommissioned, will continue storing RRSNF for decades. Consequently, safe, secure, reliable and economic storage of RRSNF will be the preferred option and will remain a crucial issue for most of the States with research reactors.

From the technical point of view and considering present available technology, there are only two options for the interim storage of RRSNF: wet storage or dry storage. Both wet and dry techniques have been used for many decades at different sites, with excellent results. Aluminium clad RRSNF has been kept intact in wet storage, with properly maintained water quality, for more than 60 years. Dry storage of RRSNF is also a fully demonstrated technology that ensures, when properly implemented, long term integrity of the RRSNF.

To ensure the integrity of the fuel elements under wet storage and their safe retrieval in the future, it is necessary to implement and maintain a comprehensive water quality programme. The programme includes corrosion monitoring and provisions to avoid deposits of settled solids on the surface of the aluminium clad, which is a potential source of pitting corrosion of the cladding. Adopting the wet option for interim storage likely costs more to operate compared to dry storage. Additionally, wet storage generates waste, mainly resin beds, which will depend on the size of the pool and the amount of spent fuel stored in it. In wet storage, process systems at the storage site have to be kept fully functional, and activities like maintenance and periodic tests and inspections have to be continuously performed and documented to guarantee the integrity and safe retrievability of the spent fuel.

In contrast to wet storage, dry storage requires less maintenance, especially if fuel is stored in sealed canisters. Key elements of dry storage include spent fuel characterization, loading, drying and encapsulation and the subsequent monitoring and surveillance. A disadvantage of interim dry storage using canisters, when compared to wet storage, is the investment required to implement the infrastructure for the drying and encapsulation processes of the spent fuel elements.

This publication provides good practice examples for the implementation of either of the two interim storage technologies. The information provided includes storage system design considerations, spent fuel characterization, surveillance programmes to address fuel degradation, system performance, and facility ageing issues.

Regardless of which technology is selected for interim storage, it will not represent the ultimate solution for final management of RRSNF. The need to make a final decision on the end point of the research reactor fuel cycle will remain a national responsibility, regardless of how long the interim storage process is drawn out.

REFERENCES

[1] INTERNATIONAL ATOMIC ENERGY AGENCY, Procedures and Techniques for the Management of Experimental Fuels from Research and Test Reactors, IAEA-TECDOC-1080, IAEA, Vienna (1999).

[2] Joint Convention on the Safety of Spent Fuel Management and on the Safety of Radioactive Waste Management, INFCIRC/546, IAEA, Vienna (1997).

[3] TOZSER, S., ADELFANG, P., BRADLEY, E., "Ten years of IAEA cooperation with the Russian research reactor fuel return programme", Reduced Enrichment for Research and Test Reactors (Proc. 33rd Int. Mtg Santiago, 2011), Argonne National Laboratory, Argonne, IL (2011).

[4] INTERNATIONAL ATOMIC ENERGY AGENCY, Experience with Spent Fuel Storage at Research and Test Reactors, IAEA-TECDOC-786, IAEA, Vienna (1995).

[5] INTERNATIONAL ATOMIC ENERGY AGENCY, Durability of Spent Nuclear Fuels and Facility Components in Wet Storage, IAEA-TECDOC-1012, IAEA, Vienna (1998).

[6] INTERNATIONAL ATOMIC ENERGY AGENCY, Understanding and Managing Ageing of Material in Spent Fuel Storage Facilities, Technical Reports Series No. 443, IAEA, Vienna (2006).

[7] INTERNATIONAL ATOMIC ENERGY AGENCY, Corrosion of Research Reactor Aluminium Clad Spent Fuel in Water, IAEA-TECDOC-1637, IAEA, Vienna (2009).

[8] INTERNATIONAL ATOMIC ENERGY AGENCY, Spent Fuel Management Options for Research Reactors in Latin America, IAEA-TECDOC-1508, IAEA, Vienna (2006).

[9] INTERNATIONAL ATOMIC ENERGY AGENCY, Return of Research Reactor Spent Fuel to the Country of Origin: Requirements for Technical and Administrative Preparations and National Experiences, IAEA-TECDOC-1593, IAEA, Vienna (2008).

[10] INTERNATIONAL ATOMIC ENERGY AGENCY, Research Reactor Spent Fuel Management: Options and Support to Decision Making, IAEA Nuclear Energy Series No. NF-T-3.9, IAEA, Vienna (2021).

[11] INTERNATIONAL ATOMIC ENERGY AGENCY, Experience of Shipping Russian-origin Research Reactor Spent Fuel to the Russian Federation, IAEA-TECDOC-1632, IAEA, Vienna (2009).

[12] INTERNATIONAL ATOMIC ENERGY AGENCY, Available Reprocessing and Recycling Services for Research Reactor Spent Nuclear Fuel, IAEA Nuclear Energy Series No. NW-T-1.11, IAEA, Vienna (2017).

[13] INTERNATIONAL ATOMIC ENERGY AGENCY, Spent Fuel Reprocessing Options, IAEA-TECDOC-1587, IAEA, Vienna (2008).

[14] INTERNATIONAL ATOMIC ENERGY AGENCY, Management and Storage of Research Reactor Spent Nuclear Fuel: Proceedings of a Technical Meeting held in Thurso, United Kingdom, 19–22 October 2009, IAEA, Vienna (2013).

[15] INTERNATIONAL ATOMIC ENERGY AGENCY, Research Reactor Database, IAEA, Vienna, https://nucleus.iaea.org/RRDB

[16] RAMANATHAN, L., HADDAD, R.A., ADELFANG, P., "A corrosion monitoring programme for research reactor spent fuel basins", Research Reactors: Safe Management and Effective Utilization: Proceedings of an International Conference Held in Sydney, Australia, 5–9 November 2007, IAEA, Vienna (2008), CD-ROM.

[17] INTERNATIONAL ATOMIC ENERGY AGENCY, Status and Trends in Spent Fuel and Radioactive Waste Management, IAEA Nuclear Energy Series No. NW-T-1.14 (Rev. 1), IAEA, Vienna (2022).

[18] INTERNATIONAL ATOMIC ENERGY AGENCY, Good Practices for Water Quality Management in Research Reactors and Spent Fuel Storage Facilities, IAEA Nuclear Energy Series No. NP-T-5.2, IAEA, Vienna (2011).

[19] DUNCAN, A.J., BANDYOPADHYAY, R.L., OLSON, C.E., L-Basin Life Expectancy, Rep. WSRC-TR-2008-00202, Westinghouse Savannah River Company, Aiken, SC (2008).

[20] INTERNATIONAL ATOMIC ENERGY AGENCY, Storage of Spent Nuclear Fuel, IAEA Safety Standards Series No. SSG-15 (Rev. 1), IAEA, Vienna (2020).

[21] PERES, M.W., "A comparison of challenges associated with sludge removal, treatment and disposal at several spent fuel storage locations", Waste Management 2007 Symposium (Proc. Int. Conf. Tucson, 2007) WM Symposia, Tucson, AZ (2007).

[22] KNIGHTEN, R., "Cleaning and draining a spent nuclear fuel pool" paper presented at DOE 12th Annual Facility Representatives Workshop, Las Vegas, NV, 2005.

[23] INTERNATIONAL ATOMIC ENERGY AGENCY, Further Analysis of Extended Storage of Spent Fuel, IAEA-TECDOC-944, IAEA, Vienna (1997).

[24] AMERICAN SOCIETY FOR TESTING AND MATERIALS, Standard Guide for Use of Protective Coating Standards in Nuclear Power Plants, ASTM D-5144-08, ASTM International, West Conshohocken, PA (2010).

[25] HOOKHAM, C., In-service Inspection Guidelines for Concrete Structures in Nuclear Power Plants, Rep. ORNL/NRC/LTR-95/14, Oak Ridge National Laboratory, Oak Ridge, TN (1995).

[26] AMERICAN SOCIETY FOR TESTING AND MATERIALS, Standard Guide for the Selection of a Leak Testing Method, ASTM E 432-91, ASTM International, West Conshohocken, PA (2011).

[27] AMERICAN SOCIETY FOR TESTING AND MATERIALS, Standard Practice for Measuring Thickness by Manual Ultrasonic Pulse-echo Contact Method, ASTM E-797-05, ASTM International, West Conshohocken, PA (2010).

[28] AMERICAN SOCIETY OF MECHANICAL ENGINEERS, 2010 ASME Boiler and Pressure Vessel Code, Section V: Nondestructive Examination, ASME, New York, NY (2010).

[29] MAZON, R., "Corrosion problems in the aluminum tank of the reactor of Mexico", GKSS--95/E/51, paper presented at Int. Topical Sem. on Management of Ageing of Research Reactors, Hamburg (1995).

[30] DIEN, N.N., TUAN, N.M., SU, T.C., "Ageing management and preventive measures for reactor pool liner and beam tubes at the Dalat research reactor", International Group on Research Reactors (IGORR) and IAEA Technical Meeting on Research Reactor Ageing, Modernisation and Refurbishment (Proc. Joint 15th Mtg Daejon, 2013), IGORR, Paris (2013).

[31] CHOWDHURY, A.Z., et al., Beam port leakage problem in the BAEC TRIGA Mark-II research reactor and the corrective measures implemented, Intl. J. Sci. Eng. Res. 4 (2013) 379–381.

[32] SYARIP, et al., "Root cause analysis of swelling problem in Kartini reactor", Research Reactors: Safe Management and Effective Utilization: Proceedings of an International Conference Held in Sydney, Australia, 5–9 November 2007, IAEA, Vienna (2008), CD-ROM.

[33] RADIOACTIVE WASTE MANAGEMENT GROUP, FORUM FOR NUCLEAR COOPERATION IN ASIA, Activity Report of Decommissioning and Clearance Task Group, Rep. FNCA RWM-005, Radioactive Waste Management Group, Tokyo (2008).

[34] LEOPANDO, L.S., "Overview of PRR-1 Decommissioning Planning", paper presented at IAEA R^2D^2 Project Workshop on Characterization Survey, Manila, 2007.

[35] ASUNCION-ASTRONOMO, A., OLIVARES, R.U., ROMALLOSA, K.M.D., MARQUEZ, J.M., "Utilizing the Philippine Research Reactor-1 TRIGA fuel in a subcritical assembly", Research Reactors: Addressing Challenges and Opportunities to Ensure Effectiveness and Sustainability, Summary of an International Conference Held in Buenos Aires, Argentina, 25–29 November 2019, IAEA, Vienna (2020).

[36] INTERNATIONAL ATOMIC ENERGY AGENCY, Survey of Wet and Dry Spent Fuel Storage, IAEA-TECDOC-1100, IAEA, Vienna (1999).

[37] AMERICAN CONCRETE INDUSTRY, Evaluation of Existing Nuclear Safety Related Concrete Structures, Rep. 349.3R-02 (R2010), ACI, Farmington Hills, MI (2010).

[38] NAUS, D.J., et al., An overview of the ORNL/NRC program to adress ageing of concrete structures in nuclear power plants, Nucl. Eng. Des. 142 (1993) 327–339.

[39] INTERNATIONAL ATOMIC ENERGY AGENCY, Core Management and Fuel Handling for Research Reactors, IAEA Safety Standards Series No. NS-G-4.3, IAEA, Vienna (2008).

[40] TRIPP, J., ARCHIBALD, K., PHILLIPS, A.M., CAMPBELL, J., "Underwater coatings for contamination control", Waste Management 2005 Symposium (Proc. Int. Conf. Tucson, 2005) WM Symposia, Tucson, AZ (2005).

[41] SINDELAR, R.L., Decay Heat Characterization of SRS Research Reactor Fuels, Rep. WSCR-TR-98-00116, Westinghouse Savannah River Company, Aiken, SC (1998).

[42] AMERICAN SOCIETY FOR TESTING AND MATERIALS, Standard Guide for Drying Behavior of Spent Nuclear Fuel, ASTM C1553-08, ASTM International, West Conshohocken, PA (2008).

[43] SINDELAR, R.L., et al., Acceptance Criteria for Interim Dry Storage of Aluminum-Alloy Clad Spent Nuclear Fuels (U), Rep. WSRC-TR-95-0347 (U), Westinghouse Savannah River Company, Aiken, SC (1996).

[44] KUHN, W., et al., Technical Review of the Characteristics of Spent Nuclear Fuel Scrap, Rep. PNNL-13751, Pacific Northwest National Laboratory, Richland, WA (2001).

[45] HURT, W.L., "Material interactions on canister integrity during storage and transport", Management and Storage of Research Reactor Spent Nuclear Fuel: Proceedings of a Technical Meeting held in Thurso, United Kingdom, 19–22 October 2009, IAEA, Vienna (2013).

[46] NUCLEAR REGULATORY COMMISSION, Standard Review Plan for Spent Fuel Dry Storage Systems at a General License Facility, Rep. NUREG-1536, Rev. 1, USNRC, Washington, DC (2010).

[47] LARGE, W., Review of Drying Methods for Spent Nuclear Fuel, Rep. WSRC-TR-97-0075, Westinghouse Savannah River Company, Aiken, SC (1997).

[48] VERBERG, M., CODÉE, H.D.K., "High level waste and spent fuel storage in the Netherlands: 10 years positive experience, extension planned", Research Reactor Fuel Management (Trans. 17th Int. Topical Mtg, St. Petersburg 2013), European Nuclear Society, Brussels (2013).

[49] SALMENHAARA, S.E.J., AUTERINEN, I., "Spent fuel management and storage at the Finnish FiR TRIGA reactor", Management and Storage of Research Reactor Spent Nuclear Fuel: Proceedings of a Technical Meeting held in Thurso, United Kingdom, 19–22 October 2009, IAEA, Vienna (2013).

[50] OBERLÄNDER, B.C., WETHE, P.I., BENNETT, P., "Storage of research reactor spent fuel in Norway", Management and Storage of Research Reactor Spent Nuclear Fuel: Proceedings of a Technical Meeting held in Thurso, United Kingdom, 19–22 October 2009, IAEA, Vienna (2013).

[51] LIAN, J.W., CHAPMAN, R.W., "Dry storage of spent fuel discharged from research reactors in Canada", Management and Storage of Research Reactor Spent Nuclear Fuel: Proceedings of a Technical Meeting held in Thurso, United Kingdom, 19–22 October 2009, IAEA, Vienna (2013).

[52] PALOMBA, M., ROSA, R., "The management of TRIGA spent fuel at ENEA RC-1 research reactor", Management and Storage of Research Reactor Spent Nuclear Fuel: Proceedings of a Technical Meeting held in Thurso, United Kingdom, 19–22 October 2009, IAEA, Vienna (2013).

[53] DIMITROVSKI, L., ANDERSON, M., "Forty-nine years of safe storage of research reactor spent fuel at ANSTO", Management and Storage of Research Reactor Spent Nuclear Fuel: Proceedings of a Technical Meeting held in Thurso, United Kingdom, 19–22 October 2009, IAEA, Vienna (2013).

[54] RÖDER, M., "Good practice of interim storage of RRSNF inside CASTOR MTR-2 flasks in Ahaus, Germany", Management and Storage of Research Reactor Spent Nuclear Fuel: Proceedings of a Technical Meeting held in Thurso, United Kingdom, 19–22 October 2009, IAEA, Vienna (2013).

[55] KASTELEIN, J., "Research reactor spent nuclear fuel, national practice for interim storage in the Netherlands", Management and Storage of Research Reactor Spent Nuclear Fuel: Proceedings of a Technical Meeting held in Thurso, United Kingdom, 19–22 October 2009, IAEA, Vienna (2013).

[56] DUIGNAN, M.R., "Experiences in the dry storage of aluminum-clad spent nuclear fuels", DOE Spent Nuclear Fuel: Challenges and Initiatives (Proc. ANS Topical Mtg Salt Lake City, 1994) Westinghouse Savannah River Company, Aiken, SC (1994).

[57] MOURÃO, R., "Dual purpose cask for dry storage of research reactor spent fuel in Latin America", Management and Storage of Research Reactor Spent Nuclear Fuel: Proceedings of a Technical Meeting held in Thurso, United Kingdom, 19–22 October 2009, IAEA, Vienna (2013).

[58] INTERNATIONAL ATOMIC ENERGY AGENCY, Costing of Spent Nuclear Fuel Storage, IAEA Nuclear Energy Series No. NF-T-3.5, IAEA, Vienna (2009).

[59] DEPARTMENT OF ENERGY, Final Environmental Impact Statement on a Proposed Weapons Nonproliferation Policy Concerning Foreign Research Reactor Spent Nuclear Fuel, Rep. DOE/EIS-0218F, Volume 2, Appendix F: Description and Impacts of Storage Technology Alternatives, USDOE, Washington, DC (1996).

[60] IYER, N.C., et al., "Interim storage and long-term disposal of research reactor spent fuel in the United States", NATO Science for Peace and Security Series, Sub-Series C: Environmental Security (LAMBERT, J., KADYRZHANOV, K., Eds), Springer, Dordrecht (2007).

[61] INTERNATIONAL ATOMIC ENERGY AGENCY, Regulations for the Safe Transport of Radioactive Material, 2018 Edition, IAEA Safety Standards Series No. SSR-6 (Rev. 1), IAEA, Vienna (2018).

[62] ELECTRIC POWER RESEARCH INSTITUTE, Industry Spent Fuel Storage Handbook, Rep. 1021048, EPRI, Palo Alto, CA (2010).

[63] INTERNATIONAL ATOMIC ENERGY AGENCY, Criticality Safety in the Handling of Fissile Material, IAEA Safety Standards Series No. SSG-27, IAEA, Vienna (2014).

[64] SINDELAR, R.L., et al., "Technology base for direct/codisposal of DOE aluminum-based SNF", DOE Spent Nuclear Fuel and Fissile Materials Management (Proc. ANS Topical Mtg Charleston, 1998), USDOE, Washington, DC (1998).

[65] SINDELAR, R.L., et al., "Evaluation of degradation during interim dry storage of aluminum-clad fuels", High Level Radioactive Waste Management (Proc. 6th Annual Int. Conf. Las Vegas, 1995), Westinghouse Savannah River Company, Aiken, SC (1995).

[66] WERTSCHING, A.K., HILL, T.J., MACKAY, N., BIRK, S.M., Material Interactions on Canister Integrity During Storage and Transport, Rep. DOE/SNF/REP-104, Idaho National Laboratory, Idaho Falls, ID (2007).

[67] SOLBRIG, C.W., KRSUL, J.R., OLSEN, D.N., "Pyrophoricity of uranium in long-term storage environments", DOE Spent Nuclear Fuel: Challenges and Initiatives (Proc. ANS Topical Mtg Salt Lake City, 1994), Westinghouse Savannah River Company, Aiken, SC (1994).

[68] AMERICAN SOCIETY FOR TESTING AND MATERIALS, Standard Guide for Evaluation of Materials Used in Extended Service of Interim Spent Nuclear Fuel Dry Storage System, ASTM C1562-03, ASTM International, West Conshohocken, PA (2003).

[69] KNOLL, R.W., GILBERT, E.R., Evaluation of Cover Gas Impurities and their Effects on the Dry Storage of LWR Spent Fuel, Rep. PNL-6365, DE88 003983, Pacific Northwest Laboratory, Richland, WA (1987).

[70] NUCLEAR REGULATORY COMMISSION, Licensing Requirements for the Independent Storage of Spent Nuclear Fuel and High-Level Radioactive Waste, and Reactor-Related Greater than Class C Waste, 10 CFR Part 72, US Gov Printing Office, Washington, DC (2021).

[71] INTERNATIONAL ATOMIC ENERGY AGENCY, Safeguards Techniques and Equipment: 2011 Edition, International Nuclear Verification Series No. 1 (Rev. 2), IAEA, Vienna (2011).

[72] INTERNATIONAL ATOMIC ENERGY AGENCY, Corrosion of Research Reactor Aluminium Clad Spent Fuel in Water, Technical Reports Series No. 418, IAEA, Vienna (2003).

[73] CHOPRA, O.K., DIERCKS, D.R., FABIAN, R.R., HAN, Z.H., LIU, Y.Y., Managing Aging Effects on Dry Cask Storage Systems for Extended Long-Term Storage and Transportation of Used Fuel (Rev. 2), Rep. FCRD-UFD-2014-000476 ANL-13/15, Argonne National Laboratory, Argonne, IL (2014).

[74] INTERNATIONAL ATOMIC ENERGY AGENCY, Data Requirements and Maintenance of Records for Spent Fuel Management: A Review, IAEA-TECDOC-1519, IAEA, Vienna (2006).

[75] BOLSHINSKY, I., THOMAS, J., CHAKROV, P., NAKIPOV, D., "The shipment of Russian-origin highly enriched uranium research reactor spent nuclear fuel from Kazakhstan", paper presented at IAEA Regional Workshop on Russian Research Reactor Fuel Return Programme Lessons Learned, Varna 2009.

[76] KUATBEKOV, R., et al., "Reduced enrichment for research reactors: Status and prospects", Research Reactors: Safe Management and Effective Utilization, Proceedings of an International Conference Held in Rabat, Morocco, 14–18 November 2011, IAEA, Vienna (2012) CD-ROM.

[77] BOESSERT, W., "Rossendorf fuel return", paper presented at IAEA Regional Workshop on Russian Research Reactor Fuel Return Programme Lessons Learned, Varna 2009.

[78] AFANASIEV, V.L., et al., "The Novosibirsk chemical concentrates plant as manufacturer of nuclear fuel and fissile materials for research reactors", Research Reactor Fuel Management (Trans. 1st Int. Topical Mtg Bruges, 1997), European Nuclear Society, Berne (1997).

[79] BOŽIČ, M., ŽAGAR., T., RAVNIK, M., "Calculation of isotopic composition during continuous irradiation and subsequent decay in biological shield of the TRIGA Mark II reactor", paper presented at Int. Conf. on Nuclear Energy for New Europe 2002, Kranjska Gora, 2002.

[80] POND, R.B., MATOS, J.E., Nuclear Mass Inventory, Photon Dose Rate and Thermal Decay Heat of Spent Research Reactor Fuel Assemblies (Rev. 1), Rep. ANL/RERTR/TM-26, Argonne National Laboratory, Argonne, IL (1996).

[81] TRELLUE., H., POSTON, D., User's Manual, Version 2.0 for Monteburns, Version 5B, Rep. LA-UR-99-4999, Los Alamos National Laboratory, Los Alamos, NM (1999).

[82] INTERNATIONAL ATOMIC ENERGY AGENCY, Post-irradiation Examination Techniques for Research Reactor Fuels, IAEA Nuclear Energy Series No. NF-T-2.6, IAEA, Vienna (in press).

[83] LEENAERS, A., VAN DEN BERGHE, S., "Microstructure of 50 year old SCK-CEN BR1 research reactor fuel", Reduced Enrichment for Research and Test Reactors (Proc. 29th Int. Mtg, Prague, 2007), Argonne National Laboratory, Argonne, IL (2007).

[84] GAN, J., et al., "Characterization of an irradiated RERTR-7 fuel plate using transmission electron microscopy", Research Reactor Fuel Management (Trans. 14th Int. Topical Mtg Marrakech, 2010), European Nuclear Society, Brussels (2010).

[85] SEARS, D.F., WANG, N., "Research Reactor Fuel Development at AECL", Reduced Enrichment for Research and Test Reactors (Proc. 23rd Int. Mtg, Las Vegas, 2000), Argonne National Laboratory, Argonne, IL (2000).

[86] KALCHEVA, S., KOONEN, E., "MCNPX 2.6.C vs. MCNPX and ORIGEN-S: state of the art for reactor core management", Research Reactor Fuel Management and Meeting of the International Group on Research Reactors (Trans. 11th Int. Topical Mtg, Lyon, 2007), European Nuclear Society, Brussels (2007).

[87] ROQUE, B., SANTAMARINA, A., "Experimental validation of actinide and fission products inventory from chemical assays in French PWR spent fuels", Practices and Developments in Spent Fuel Burnup Credit Applications, IAEA-TECDOC-1378, IAEA, Vienna (2003).

[88] LEBENHAFT, J.R., MACIAN-JUAN, R., "Deterministic and stochastic analysis of the isotopic composition of highly burned MOX fuel", Paul Scherrer Institut Scientific Report 2003, Volume IV: Nuclear Energy and Safety, Paul Scherrer Institut, Villigen (2004).

[89] RADULESCU, G., WAGNER, J.C., "Review of results for the OECD/NEA phase VII benchmark: study of spent fuel compositions for long-term disposal", paper presented at Int. Conf. on High-Level Radioactive Waste Management, Albuquerque 2011.

[90] INTERNATIONAL ATOMIC ENERGY AGENCY, Safety of Nuclear Fuel Cycle Facilities, IAEA Safety Standards Series No. SSR-4, IAEA, Vienna (2017).

[91] INTERNATIONAL ATOMIC ENERGY AGENCY, Predisposal Management of Radioactive Waste, IAEA Safety Standards Series No. GSR Part 5, IAEA, Vienna (2009).

[92] INTERNATIONAL ATOMIC ENERGY AGENCY, Safety Assessment for Facilities and Activities, IAEA Safety Standards Series No. GSR Part 4 (Rev. 1), IAEA, Vienna (2016).

[93] INTERNATIONAL ATOMIC ENERGY AGENCY, Leadership and Management for Safety, IAEA Safety Standards Series No. GSR Part 2, IAEA, Vienna (2016).

[94] EUROPEAN ATOMIC ENERGY COMMUNITY, FOOD AND AGRICULTURE ORGANIZATION OF THE UNITED NATIONS, INTERNATIONAL ATOMIC ENERGY AGENCY, INTERNATIONAL LABOUR ORGANIZATION, INTERNATIONAL MARITIME ORGANIZATION, OECD NUCLEAR ENERGY AGENCY, PAN AMERICAN HEALTH ORGANIZATION, UNITED NATIONS ENVIRONMENT PROGRAMME, WORLD HEALTH ORGANIZATION, Fundamental Safety Principles, IAEA Safety Standards Series No. SF-1, IAEA, Vienna (2006).

[95] HAAPALEHTO, T., WILMER, P., "Spent fuel storage, a long term engagement: OECD/NEA overview", Storage of Spent Fuel from Power Reactors, C&S Papers Series 20, IAEA, Vienna (2003).

[96] INTERNATIONAL ATOMIC ENERGY AGENCY, Site Evaluation for Nuclear Installations, IAEA Safety Standards Series No. SSR-1, IAEA, Vienna (2019).

[97] The Convention on the Physical Protection of Nuclear Material, INFCIRC/274/Rev.1, IAEA, Vienna (1980).

[98] INTERNATIONAL ATOMIC ENERGY AGENCY, Nuclear Security Recommendations on Physical Protection of Nuclear Material and Nuclear Facilities (INFCIRC/225/Revision 5), IAEA Nuclear Security Series No. 13, IAEA, Vienna (2011).

[99] Amendment to the Convention on the Physical Protection of Nuclear Material, INFCIRC/274/Rev.1/Mod.1 (Corrected), IAEA, Vienna (2021).

[100] Treaty on the Non-Proliferation of Nuclear Weapons, INFCIRC/140, IAEA, Vienna (1970).

[101] ABEDIN-ZADEH, R., BOSLER, G., CARCHON, R., LEBRUN, A., "IAEA safeguards verification methods for spent fuel in wet and dry storage", Storage of Spent Fuel from Power Reactors, C&S Papers Series 20, IAEA, Vienna (2003).

[102] INTERNATIONAL ATOMIC ENERGY AGENCY, Decommissioning of Nuclear Power Plants, Research Reactors and Other Nuclear Fuel Cycle Facilities, IAEA Safety Standards Series No. SSG-47, IAEA, Vienna (2018).

[103] FOOD AND AGRICULTURE ORGANIZATION OF THE UNITED NATIONS, INTERNATIONAL ATOMIC ENERGY AGENCY, INTERNATIONAL CIVIL AVIATION ORGANIZATION, INTERNATIONAL LABOUR ORGANIZATION, INTERNATIONAL MARITIME ORGANIZATION, INTERPOL, OECD NUCLEAR ENERGY AGENCY, PAN AMERICAN HEALTH ORGANIZATION, PREPARATORY COMMISSION FOR THE COMPREHENSIVE NUCLEAR-TEST-BAN TREATY ORGANIZATION, UNITED NATIONS ENVIRONMENT PROGRAMME, UNITED NATIONS OFFICE FOR THE COORDINATION OF HUMANITARIAN AFFAIRS, WORLD HEALTH ORGANIZATION, WORLD METEOROLOGICAL ORGANIZATION, Preparedness and Response for a Nuclear or Radiological Emergency, IAEA Safety Standards Series No. GSR Part 7, IAEA, Vienna (2015).

[104] FOOD AND AGRICULTURE ORGANIZATION OF THE UNITED NATIONS, INTERNATIONAL ATOMIC ENERGY AGENCY, INTERNATIONAL LABOUR OFFICE, PAN AMERICAN HEALTH ORGANIZATION, UNITED NATIONS OFFICE FOR THE COORDINATION OF HUMANITARIAN AFFAIRS, WORLD HEALTH ORGANIZATION, Arrangements for Preparedness for a Nuclear or Radiological Emergency, IAEA Safety Standards Series No. GS-G-2.1, IAEA, Vienna (2007).

[105] INTERNATIONAL ATOMIC ENERGY AGENCY, Cost Estimation for Research Reactor Decommissioning, IAEA Nuclear Energy Series No. NW-T-2.4, IAEA, Vienna (2013).

[106] SINDELAR, R.L., DEIBLE, R.W., Demonstration of Long-Term Storage Capability for Spent Nuclear Fuel in L Basin, Rep. SRNL-STI-2011-00190, Savannah River National Laboratory, Aiken, SC (2011).

[107] MAXTED, M., "Overview and Status Update of the Savannah River Site Spent Nuclear Fuel Program", paper presented at Savannah River Site Citizens Advisory Board on Nuclear Materials Committee, Savannah, GA, 2018.
[108] BURKE, S.D., HOWELL, J.P., "The impacts of prolonged wet storage of DOE reactor irradiated nuclear materials at the Savannah River Site", DOE Spent Nuclear Fuel: Challenges and Initiatives (Proc. ANS Topical Mtg Salt Lake City, 1994), Westinghouse Savannah River Company, Aiken, SC (1994).
[109] HOWELL, J.P., ZAPP, P.E., NELSON, D.Z., "Corrosion of aluminum alloys in a reactor disassembly basin", NACE-93609, paper presented at 48th NACE Annual Conference and Corrosion Show 1993, New Orleans, LA, 1993.
[110] ROSE, D., "Spent fuel management at Savannah River Site", paper presented at Institute of Nuclear Materials Management (INMM) 28th Spent Fuel Management Seminar, Arlington, VA, 2013.
[111] GILLAS, D., "SRS used nuclear fuel management", paper presented at Savannah River Site Citizens Advisory Board on Nuclear Materials Committee, Martinez, CA, 2011.
[112] BATES, B., "SRS used nuclear fuel management", paper presented at Savannah River Site Citizens Advisory Board on Nuclear Materials Committee, Aiken, SC, 2011.
[113] ROSE, D., "Used nuclear fuel management at Savannah River Site (SRS)", paper presented at Institute of Nuclear Materials Management (INMM) 27th Spent Fuel Management Seminar, Arlington, VA, 2012.
[114] ROSE, D., "Used nuclear fuel receipt, shipping and disposition at Savannah River Site (SRS)", paper presented at Institute of Nuclear Materials Management (INMM) 27th Spent Fuel Management Seminar, Arlington, VA, 2014.
[115] VINSON, D.W., et al., "US practice for interim wet storage of RRSNF", Management and Storage of Research Reactor Spent Nuclear Fuel: Proceedings of a Technical Meeting held in Thurso, United Kingdom, 19–22 October 2009, IAEA, Vienna (2013).
[116] HATHCOCK, D.J., et al., "Spent nuclear storage basin water chemistry: electrochemical evaluation of aluminum corrosion", paper presented at NACE Int. Corrosion Conf. 2008, New Orleans, 2008.
[117] CARLSEN, B., et al., "Experience With Damaged Spent Nuclear Fuel at U.S. DOE Facilities", Paper No. ICONE 14-89319, presented at 14th Int. Conf. on Nuclear Engineering, Miami, 2006.
[118] HOWELL, J.P., "Corrosion surveillance in spent fuel storage pools", NACE-97107, paper presented at Corrosion/97 Research Topical Symp. New Orleans, 1997.
[119] VORMELKER, P.R., MERCADO, D., DUNCAN, A.J., "Corrosion surveillance of aluminum alloys in spent fuel storage basin", NACE-05595, paper presented at NACE Int. Corrosion Conf. 2005, Houston, 2005.
[120] MINISTRY OF INFRASTRUCTURE AND WATER MANAGEMENT, Joint Convention on the Safety of Spent Fuel Management and on the Safety of Radioactive Waste Management, National Report of the Kingdom of the Netherlands for the Seventh Review Meeting, Ministry of Infrastructure and Water Management, The Hague (2020).
[121] KASTELEIN, J., CODÉE, H.D.K., "HABOG: One building for all high level waste and spent fuel in the Netherlands. The first year of experience", Research Reactor Fuel Management (Trans. 9th Int. Topical Mtg Budapest, 2005), European Nuclear Society, Brussels (2005) 174–180.
[122] DE VOS, R.M., ROOBOL, L.P., SCHMALZ, F., "Semi-long term storage of uranium containing waste from molybdenum production" paper presented at Hotlab Conf., Jülich, 2006.
[123] TOZSER, S., "Wet and semi-dry storage of spent nuclear fuel at the Budapest research reactor", Management and Storage of Research Reactor Spent Nuclear Fuel: Proceedings of a Technical Meeting held in Thurso, United Kingdom, 19–22 October 2009, IAEA, Vienna (2013).
[124] HARGITAI, T., VIDOVSZKY, I., "New storage mode for spent fuel at the Budapest research reactor", Research Reactor Fuel Management (Trans. 6th Int. Topical Mtg Ghent, 2002), European Nuclear Society, Berne (2002) 109–113.
[125] TOZSER, S., HARGITAI, T., VIDOVSZKY, I., "Encapsulation of nuclear spent fuel for semi-dry storage at the Budapest research reactor", Research Reactor Utilization, Safety Decommissioning, Fuel and Waste Management: Proceedings of an International Conference 10–14 November 2003 Santiago, Chile, IAEA, Vienna (2005).
[126] MATAUŠEK, M.V., MARINKOVIĆ, N.M., VUKADIN, Z., "Research Reactor RA at the Vinča Institute of Nuclear Sciences. Ageing, Refurbishment and Irradiated Fuel Storage", Research Facilities for the Future of Nuclear Energy, World Scientific Publishing, Singapore (1996).
[127] PEŠIĆ, M., KOLUNDŽIJA, V., LJUBENOV, V., CUPAĆ, S., "Causes of extended shutdown state of "RA" research reactor in "Vinča" institute", RER/9/058, paper presented at the IAEA Regional Workshop on Extended Shutdown and Decommissioning of Research Reactors, Riga, 2001.

[128] PEŠIĆ, M., et al., "Vinča Nuclear Decommissioning Program", Society for Electronic, Telecommunications, Computers, Automation and Nuclear Engineering (Proc. 46th ETRAN Conf. Banja Vrućica, 2002), ETRAN Society, Belgrade (2002).

[129] LJUBENOV, V., PEŠIĆ, M., ŠOTIĆ, O., "RA Research Reactor in "Vinča" Institute – Approach to the Decommissioning", International Yugoslav Nuclear Society Conference (YUNSC-2002) (Proc. Int. Conf., Belgrade, 2002), VINČA Institute of Nuclear Sciences, Belgrade (2003) CD-ROM.

[130] PEŠIĆ, M., SUBOTIĆ, K., LJUBENOV, V., ŠOTIĆ, O., "Vinča Nuclear Decommissioning Program – Establishment and Initialisation", Nuclear Energy, Nuclear Power – A New Challenge (SIEN 2003) (Proc. Int. Symp. Bucharest, 2003), Romanian Nuclear Energy Association, Bucharest (2003).

[131] ŠOTIĆ, O., "Spent fuel repatriation from the Republic of Serbia", Research Reactors: Safe Management and Effective Utilization: Proceedings of an International Conference Held in Rabat, Morocco, 14–18 November 2011, IAEA, Vienna (2012) CD-ROM.

[132] PEŠIĆ, M., ŠOTIĆ, O., HOPWOOD, W.H., Transport of High Enriched Uranium Fresh Fuel from Yugoslavia to the Russian Federation, Nucl. Technol. Radiat. Prot. 17 (2002) 71–76.

[133] PEŠIĆ, M., ŠOTIĆ, O., "Experiences from HEU fresh fuel transportation from Yugoslavia to Russia", Research Reactor Fuel Management (Trans. 7th Int. Topical Mtg, Aix-en-Provence 2003), European Nuclear Society, Brussels (2003) 220–224.

[134] KNEŽEVIĆ, I., "Uranium Take-back Programme in Serbia", paper presented at IAEA Technical Meeting on Lessons Learned from High Enriched Uranium Take-back Programmes, Gdansk, 2019.

[135] THE "BORIS KIDRIČ" INSTITUTE OF NUCLEAR SCIENCES, Annual Reports of the RA Reactor Operation, Vinča Institute of Nuclear Sciences, Vinča, Yugoslavia (1960–1984).

[136] BORIS KIDRIČ INSTITUTE OF NUCLEAR SCIENCES, Study on Spent Fuel Storage Pool of the Reactor RA, Internal Report Z5.A1.4, Vinča Institute of Nuclear Sciences, Vinča, Yugoslavia (1984).

[137] ŠOTIĆ, O., et al., Research Reactor RA – Operation and Maintenance Report in 1984, Internal Report, Vinča Institute of Nuclear Sciences, Vinča, Yugoslavia (1984).

[138] INTERNATIONAL ATOMIC ENERGY AGENCY, Appendix II to IAEA Mission Report to Vinča Institute in February 1997, IAEA, Vienna (1997).

[139] PEŠIĆ, M., et al., Study of corrosion of aluminium alloys of nuclear purity in ordinary water – Part one, Nucl. Technol. Radiat. Prot. 19 (2004) 77–93.

[140] PEŠIĆ, M., et al., Study of corrosion of aluminium alloys of nuclear purity in ordinary water – Part two, Nucl. Technol. Radiat. Prot. 20 (2005) 45–60.

[141] PLECAS, I., PAVLOVIC, R., PAVLOVIC, S., Development of solidification techniques for radioactive sludge produced by a research reactor, Prog. Nucl. Energy 44 (2004) 43–47.

[142] PEŠIĆ, R., et al., "Radioactive Waste Management in Serbia, 2002–2010", Nuclear Energy for New Europe 2011 (Proc. 20th Int. Conf. Bovec 2011), Nuclear Society of Slovenia, Ljubljana, 2011.

[143] ADEN, V.G., et al., "Spent fuel from RA reactor inspection of state and options for management", International Yugoslav Nuclear Society Conference (YUNSC-2002) (Proc. Int. Conf., Belgrade, 2002), VINČA Institute of Nuclear Sciences, Belgrade (2003) CD-ROM.

[144] ŠOTIĆ, O., PEŠIĆ, M., et al., Preliminary analyses of the feasibility of spent fuel management (Phase 1), Rep. Vinča-NTI-114 on IAEA TCP Contract No. SCG/4/003-89102A 'Safe Removal of Spent Fuel of the Vinča RA Research Reactor', Vinča, Serbia (2003).

[145] FISHER, D.L., et al., Water Chemistry Control System for Recovery of Damaged and Degraded Spent Fuel, Rep. SRNL-STI-2011-00100, Savannah River National Laboratory, Aiken, SC (2010).

[146] NAQVI, S.J., FROST, C.R., "An update on used fuel management in Canada", Spent Fuel Management: Current Status and Prospects 1993, IAEA-TECDOC-732, IAEA, Vienna (1994).

[147] SHIRAI, K., et al., Testing of Metal Cask and Concrete Cask, Management of Spent Fuel from Nuclear Power Reactors: Proceedings of an International Conference held in Vienna, 31 May–4 June 2010, IAEA, Vienna (2015) CD-ROM.

ABBREVIATIONS

AFR	away-from-reactor
ANSTO	Australian Nuclear Science and Technology Organisation
AR	at-reactor
BRR	Budapest Research Reactor
BSF	bulk shielding facility
COVRA	Central Organization for Radioactive Waste
CRP	coordinated research project
HABOG	high level waste treatment and storage building, Netherlands
HEU	high enriched uranium
HFIR	High Flux Isotope Reactor
HLW	high level waste
INL	Idaho National Laboratory
LEU	low enriched uranium
MCO	multicanister overpack
MTR	materials testing reactor
NDE	non-destructive examination
NPP	nuclear power plant
PNRI	Philippine Nuclear Research Institute
RBOF	Receiving Basin For Offsite Fuels
RRSNF	research reactor spent nuclear fuel
SNF	spent nuclear fuel
SRS	Savannah River Site
SSCs	structures, systems and components
SSCH	stainless steel channel holder
TRIGA	Training, Research, Isotopes, General Atomics
USDOE	United States Department of Energy
VTS	vertical tube storage
VVR	Vodo-Vodjanoj Reaktor

CONTRIBUTORS TO DRAFTING AND REVIEW

Adelfang, P.	International Atomic Energy Agency
Fuentes Solis, N.	International Atomic Energy Agency
Geupel, S.	International Atomic Energy Agency
Hanlon, T.	International Atomic Energy Agency
Iyer, N.C.	Savannah River National Laboratory, USA
Kastelein, J.	COVRA, Netherlands
Lian, J.W.	Chalk River Laboratories, Atomic Energy of Canada Ltd, Canada
Marshall, F.M.	International Atomic Energy Agency
Muhammad Nor, A.W.	International Atomic Energy Agency
Soares, A.J.	International Atomic Energy Agency
Standring, P.	International Atomic Energy Agency
Tőzsér, S.	International Atomic Energy Agency
Verberg, M.	COVRA, Netherlands
Vinson, D.W.	Savannah River National Laboratory, USA

Consultants Meetings

Vienna, Austria: 15–17 December 2008, 4–6 May 2011, 7–9 July 2014

Technical Meeting

Thurso, United Kingdom: 19–23 October 2009

Structure of the IAEA Nuclear Energy Series*

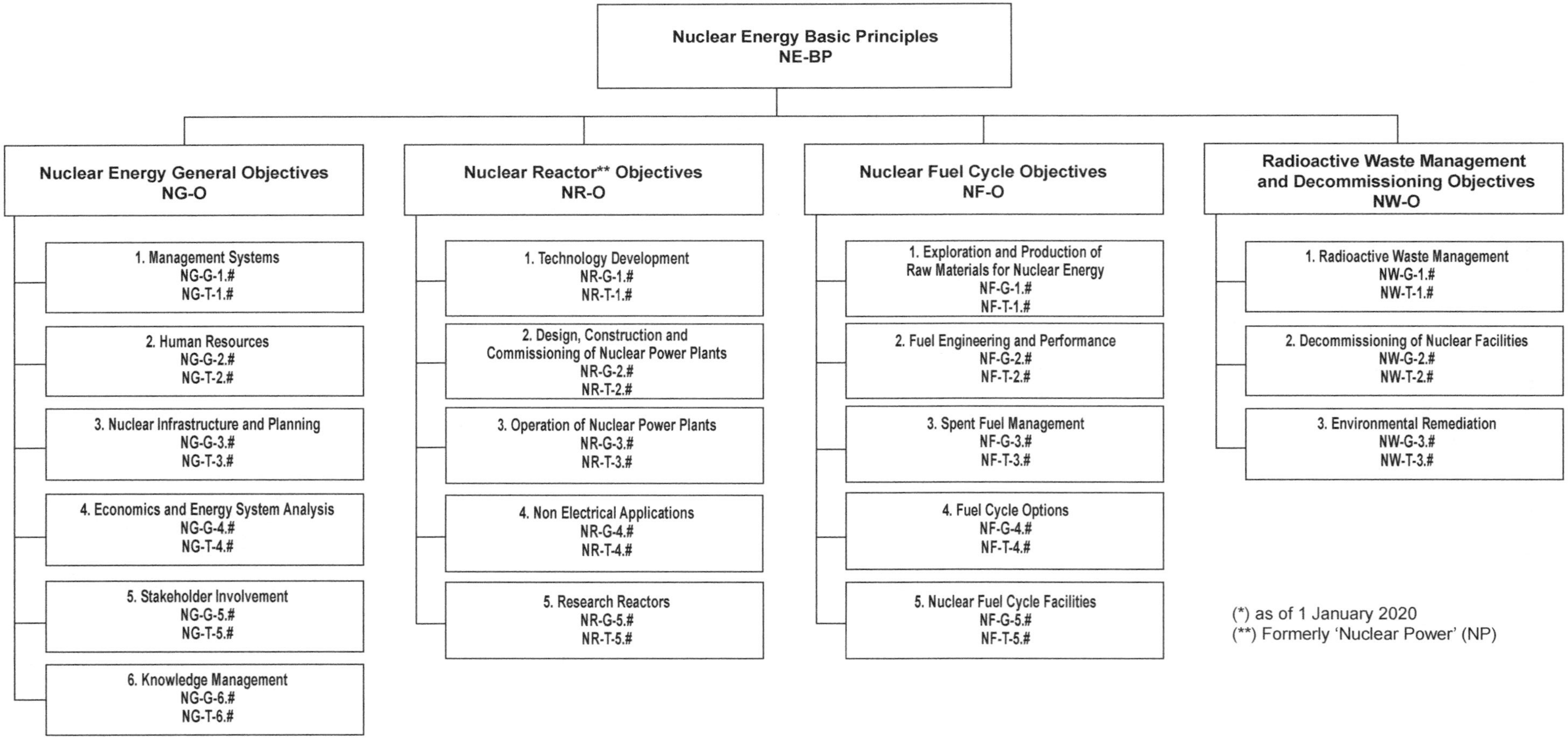

Nuclear Energy Basic Principles
NE-BP

Nuclear Energy General Objectives
NG-O

1. Management Systems
NG-G-1.#
NG-T-1.#

2. Human Resources
NG-G-2.#
NG-T-2.#

3. Nuclear Infrastructure and Planning
NG-G-3.#
NG-T-3.#

4. Economics and Energy System Analysis
NG-G-4.#
NG-T-4.#

5. Stakeholder Involvement
NG-G-5.#
NG-T-5.#

6. Knowledge Management
NG-G-6.#
NG-T-6.#

Nuclear Reactor Objectives**
NR-O

1. Technology Development
NR-G-1.#
NR-T-1.#

2. Design, Construction and Commissioning of Nuclear Power Plants
NR-G-2.#
NR-T-2.#

3. Operation of Nuclear Power Plants
NR-G-3.#
NR-T-3.#

4. Non Electrical Applications
NR-G-4.#
NR-T-4.#

5. Research Reactors
NR-G-5.#
NR-T-5.#

Nuclear Fuel Cycle Objectives
NF-O

1. Exploration and Production of Raw Materials for Nuclear Energy
NF-G-1.#
NF-T-1.#

2. Fuel Engineering and Performance
NF-G-2.#
NF-T-2.#

3. Spent Fuel Management
NF-G-3.#
NF-T-3.#

4. Fuel Cycle Options
NF-G-4.#
NF-T-4.#

5. Nuclear Fuel Cycle Facilities
NF-G-5.#
NF-T-5.#

Radioactive Waste Management and Decommissioning Objectives
NW-O

1. Radioactive Waste Management
NW-G-1.#
NW-T-1.#

2. Decommissioning of Nuclear Facilities
NW-G-2.#
NW-T-2.#

3. Environmental Remediation
NW-G-3.#
NW-T-3.#

(*) as of 1 January 2020
(**) Formerly 'Nuclear Power' (NP)

Key

BP: Basic Principles
O: Objectives
G: Guides and Methodologies
T: Technical Reports
Nos 1–6: Topic designations
#: Guide or Report number

Examples

NG-G-3.1: Nuclear Energy General (**NG**), Guides and Methodologies (**G**), Nuclear Infrastructure and Planning (topic **3**), **#1**
NR-T-5.4: Nuclear Reactors (**NR**), Technical Report (**T**), Research Reactors (topic **5**), **#4**
NF-T-3.6: Nuclear Fuel (**NF**), Technical Report (**T**), Spent Fuel Management (topic **3**), **#6**
NW-G-1.1: Radioactive Waste Management and Decommissioning (**NW**), Guides and Methodologies (**G**), Radioactive Waste Management (topic **1**) **#1**